A REFLEXIVE FOUNDATION FOR HUMAN SCIENCE

CONTENTS

CHAPTER 1

THE IDEA OF REFLEXIVE HUMAN SCIENCE
AND THE BEGINNING REALIZATIONS AND ISSUES
OF REFLEXIVIZATION

In reflexive human science,[1] the subject, that is, what experiences and intends, what 'experientions',[2] reflexivizes. The subject's reflexivization is it's experientioning of it occurring, and, through this self-experientioning, its reconstructive explication of its issue of that and what it is. In this reflexive reconstruction, from a spirit-science perspective, the subject makes what is generalizable in its being-process, the objectivity in its being-process, issue for itself. Experientioning itself occurring, the subject examines horizons including: 'that it is', 'what it does', 'how it works', and 'what it can be'.

The subject begins reflexivization, begins clarifying it's experientioning of itself occurring, in experiention-of-being, and proceeds with it by reconstructing a generalizable inner 'spirit-mechanism': it reconstructs, namely, 'unitary mechanism' (the minimal definition of human being). And, beyond this, it reconstructs that in which it is what it is generally; that is, it develops the encompassing definition of human existence to include reconstruction of narrative-themes, subsystems, issues of organizational structure, and so on. In equipping itself with this generaliseable and reproducible (scientific[3]) reconstruction that yields a definition of human existence and a 'feedback system' in its being-process, the subject is able in aspects to 'be scientifically'. It is able to 'be scientifically' because now, in pure − reflexive − self-experientioning, as a result of its reflexive practise and scientific reconstruction, it is mediated with itself using a scientific self-descriptive language − a self-knowledge that includes the laws of its being-process and knowledge of structure in its situation generally − that it can apply itself to itself with and use in its situation generally.

The idea of reflexive human science, then, is that the 'subject's reflexivization' is the practise of reflexive human science, practise in which the self-experientioning subject's awareness of that it is (of its being a being-process) develops into an examined and reproducible reflexive perspective in which it reconstructs the generalizable in what its being-process is and how it works and what it can be (in that there is an issue of what it can make itself be), and in which it explicates as well the issue of what it is the being of (the issue of Being). The subject's reflexivization, explicated here as a reflexive foundation for human science, begins, then, with experiention-of-being: in experiention-of-being the subject encounters and clarifies the issue of its being (of that and what it is); and, doing so, it encounters as well concomitantly both the issue of knowing it to be so and the issue of what is in that there is something.

Beginning with experiention-of-being, reflexivization begins with, or as, 'epistemologization-ontologization'.[4] This involves the concepts of being (existence, experientioning), knowing (a mode of experientioning), and Being (What-can-be, That which is what-is, What what-is is the Occurring of). In beginning addressing the issue there is in its experientioning of itself occurring, the reflexivizing subject demonstrates to itself that because it encounters an issue, because there is an issue for it, it must be that for which an issue is, that by which something – a what-is, a meaning, existing – is being experienced and intended, that is, in some degree understood. And it demonstrates to itself that this being so therefore entails that there is something, namely it itself, and so too That (the What-can-be, Being) which can be it. The reflexivizing subject sees that it recognizes that it is that for which an issue is because it sees that this being so is necessarily predicated in the conditions of the possibility of there being an issue for it, which it experientions is occurring, whatever else the issue is about (for example 'that a sense-datum'). And it sees that for it, in that there must be What is being it if it is (which it knows in clarifying there being an issue for it, a what-is), what is in general is What-can-be-being-what-is, is, that is, Meaning.[5]

In 'epistemologization-ontologization' and beyond the subject manages its experientioning of itself occurring and reflexive reconstruction of it scientifically, as knowledge, by validating its assertions about itself with reference to evidence it gets in experientions of itself occurring and through what it sees and demonstrates is the case surrounding that. It does this in the actual situation and idea that its experiention of itself is what it is being, is the experiention of what it is that it has to work with, and is experiention of what it is that its being-process is; that is, (and in thinking through such as 'necessary predication' and 'being in-itself' and 'the significance of conditions of its possibility' and 'what must be functioning in that its experientioning is' and so on) it does this reflexively: so, to 'know it to be so' is to encounter it in the conditions of existence and validate it with reference to evidence there in and for self-experientioning, so for example 'that an issue is' is known to be 'because it is (reliably, reproducibly, etc.) experientioned' (or, 'because there is a reliable, reproducible, etc. experiention of one'), so 'that it is' is known to be because 'there would not be an issue for it if it were not there experientioning it', so it is known that 'because there is its experientioning it has an issue of its conditions of possibility',[6] and so on.

In and through these beginning realizations, in seeing 'the structure of necessary predication' in the 'reflexive circle',[7] the subject both grounds its certainty of its existence (certainty of that it is and is what is encountering an issue) as it is possible for it to, and, will there through understanding, in experientioning what-is and oriented in awareness of being 'in-itself ' and encountering its responsibility in being 'for-itself ',[8] the subject can proceed systematically further with both the process of reconstructing and developing how it exists, and with the examination of Being (thinking through the idea that there are laws of It, the idea of that it, the reflexivizing subject, is an in-Itself of It, and so on).

CHAPTER 2

THE MINIMAL DEFINITION OF HUMAN BEING:
THE LAW OF UNITARY MECHANISM

The reflexivizing subject finds that it can proceed beyond its beginning realizations and issues. The reflexivizing subject finds that the realizations and issues in experiention-of-being soon develop to include issues of how it has come to be at all, and exploration of what is involved in and possible for it: for example, 'in evolving life', 'through understanding', 'in language', 'in embodiment', 'in an issue of knowing it to be so', 'in the world', 'in ground-relations', 'in 'the that' in it', 'in strategies for managing doing', 'in melt', 'with issues of survival and happiness, reason, power, and right', 'as something What-can-be is being Itself as', and so on. Proceeding beyond beginnings in epistemologization-ontologization as such in this reflexive way, the only way possible,[9] then, the reflexivizing subject can proceed to further, more detailed and systematic − integrative − examinations of how it exists, of 'how its being-process is', in which it can arrive at a minimal definition of human being through a reconstruction of 'unitary mechanism'.

The reflexivizing subject's reconstruction of unitary mechanism is its reconstruction of 'what it is' through a reflexive reconstructive experientioning of 'what it does'. In reconstructing unitary mechanism it sees, namely, the fact of the mechanism of its being-process, the fact that, objectively, for and in being itself, it is what exists 'producing-using meaning to make its world the world it prefers as it can make it work'. It sees this, the minimal definition of human being in the law of unitary mechanism: human beings exist producing-using meaning to make the world as preferred as can be made to work. And, pursuing the issue further still, the reflexive subject sees that this existence, rooted generally in 'necessary conditions' and 'involved in grounds', involves it in intrinsic 'conditions of participation' (in itself and with others). It sees, namely, that its being-process includes a need to participate in the five intrinsic and concomitant narrative-themes of 1) comprehension, 2) truth, 3) compatibility, 4) attitude and orientation, and 5) emotional and spiritual stability and embodiment.[10] And it sees such as that there is contingency in 'how it is and can be what it intrinsically is', that is, it sees contingency in how instances of 'unitary mechanism' are what they are, in that one or another thing could be value, in that things need to be made to work, and often can be made to work in one or another way, and the like.

All issues of instinct, substance, understanding, development, belief, desire, knowledge, reality, possibility, truth, sublimation-utilization, repression, right, revolution, aesthetic, construction, interaction, society, evolution, and so on, all issues in human existence, in being an experientioning, are apparent through and in the environment and structure of 'unitary mechanism'. Reflexively reconstructing and self-grounding itself in this way, the subject recognizes that it is a spirit-mechanism in a spirit-environment and embodiment and world (grounds) generally, and that existing in this way it has responsibility in self-regulating and being-process generally, involving self-causation and an issue of freedom, along with issues of being caused in: the reflexive subject encounters that how it is what it is (involving issues ranging from doing good/evil, being in 'the that', how it is intersubjectively connected, to what it eats and biography generally) makes a difference in what happens in it and what it can achieve, in its state, energy and health and so on.[11] In this knowledge of itself, the reflexive subject has awareness of its objectivity, of what it is (and can be) in how What-can-be-is-being-what-is, knowledge, namely, of being 'reflexive-subject-in-grounds'. And so, as well, it has awareness of What-can-be-being-what-is – ultimately the transcendent involving paradox, the Infinite – generally. And it develops awareness of its role in itself, of the relation of its responsibility in its self-formation and being to the variables and determinations of its environment, including spirit-environment.

This concept of the law of human being (of unitary mechanism), involving testable hypotheses, works to integrate various disciplines (biology, psychology, sociology, political science, economics, history, medicine, jurisprudence, evolutionary science, ecology, etc.) in one 'human science'; in, that is, the reflexive science reflexive-subject-in-grounds does of the process of the meta-system (involving life) in the operations of which the world-of-understanding is produced-used.[12] Meta-system involves a difference, the difference of a level of contingency located in the responsibility and issue of freedom (experiention of being a self-causing self-activity) in the subject: the subject is (reflexively recognizes itself to be) a subsystem-with-a-difference, namely the 'participant-competence' subsystem-with-a-difference, a subsystem-with-a-difference that occurs concomitantly along with other subsystems in the process of meta-system. The other subsystems (similarly in aspects involving a difference linked to the activities of the subject) are: society, polity, culture, and economy/ecology.[13]

Proceeding in reflexivizing, in the issue of 'how it exists', the subject sees that in its being-process, in being what it is – in producing-using meaning to make world as preferred as can be made to work, as the participant-competence (subsystem-with-a-difference) in meta-system involved in narrative-theme-mediated activity – it coordinates 'triadizings of value'. This means that it coordinates triads of its being-process in which, if it can be made to work, a possibility – a value, a preferred world – is actualized as an achieved difference. That is, this means that, through concern, a will operating through understanding, and including encounter with responsibility and the issue of freedom in it, a subject's being-process involves, successfully or otherwise, in three moments, namely, in possibility/actualisation-of-possibility/

achieved difference. The reflexive subject identifies what it coordinates triadizing about as issues of concern, and sees that because the world exists for it always already as an issue of being preferred world, meaning (that is, what-is, or, against the background of 'limit at infinity', Meaning) exists for it as value, and as such as within reach or not (reach that involves conditions such as know-how, resources, and so on). Aware of itself coordinating triadizings, of itself existing, amongst other subsystems in meta-system (namely, polity, culture, economy/ecology, society), as participant-competence subsystem-with-a-difference (involving 'conditions of participation', 'narrative themes' etc.), it identifies that issues arise for reflexive reconstruction such as of what structure is intrinsic in a paradigm of meta-system and what contingent, issues of intrinsic laws and constructions, process, differentiation and integration, telos, infinity, and so on.

Put more extensively, for subjectivity, being confronted with value and involved through concern in issues of possibility/actualization/achieved difference is to be involved, via narrative-theme-involvement, in doing 'processes of rationality'. The subject emerges and functions in a process that, at the limit of survival and in managing generally, ties it to having and giving reasons: the subject emerges and functions to be what it is in what it is what it is in in a limit of rationality. The subject must sufficiently be (put itself) in coordinated rationalities if its production-use of meaning (coordination of triadizings, being of what it is) is to work enough for survival, and, amongst the unexpected, synergic processes, and serendipity and the rest, if it is to work, in general, in making sense in one or another way. This rationality-requirement in how it works, shaped by intrinsic properties of the structure of its situation and of survival-issues, and reaching beyond these as well, is a dimension in the objectivity of subjectivity. It is a dimension of objectivity in which, for example in differentiations of culture such as science, moves such as 'citing evidence' in experiment and observation are held as reasons, just as at other moments such as 'feeling like it' can be held as a reason.[14] The rationality-requirement is a dimension in how the participant-competence subsystem-with-a-difference can function. Rationality is a dimension in the subject's being what it is in What-can-be-being-what-is, and as regards, in concern with, what truth in That is, that is, is a dimension in how it is for 'unitary mechanism' to be what it is and 'have truth – how-What-can-be-can-and-cannot-be, ultimately itself and Reason – to work with'. The subject needs to be in sufficient truth-congruence if 'reality-effects' are to be made to happen and 'will-to-power-of-freedom' (a preferred world) sufficiently fulfilled, that is, self-realized.

The general principle in human activity reconstructed in reflexive human science, then, is that the subject, defined in terms of unitary mechanism, and in terms of What-can-be-being-What-is (as something It is doing), acts to enhance its value as it can make work, to manifest freedom (realized value); and the explanation of its activities includes issues of there being, in this, a narrative-theme-structure, issues of such as the sufficient-rationality requirement, and integration and power, includes influence of characteristics, gender, the environment of organizational structure involving its systems, diet, and whatever else, in general includes conditions of existence, the relation to the difference grounds make in how triadizing occurs (including intelligence, genes, conditioning, biography, and so on). And, in this, it is clear that, as the process of reflexivization in which subjectivity recognizes its objectivity and turns itself into reflexive subjectivity develops, and knowledge of 'that it is' becomes a developed identification (reflexive reconstruction) of 'what it is', the subject's further examining of 'how it exists' works for its potential in self-determination.[15] It is clear that it works this way through the development and application of knowledge of grounds, in self-application, in regulating in and constructing 'how it is being and can be what it is'. The reflexive subject develops the potential in reflexivization, then, by increasing its knowledge and practice of objectivity in its situation, by increasing and using knowledge of that and how grounds make a difference, including in what is possible for pure reflexive subjectivity, what is possible in relation with 'the that', what is possible in allowing what can to happen or not, and so on.

Three interlinked levels of reflexive human science, of science of the process of the meta-system-with-a-difference involved in the production-use of the world-of-understanding, are apparent: 1) reconstructive (knowledge of that and what we are and how we work), 2) constructive (applications of knowledge of how we work in us being what we are), and 3) the level of explaining why we are what we are as we are, including in constructing, a level including issues of what, at any time and if anything else, is causing in one causing in and for oneself, and so on. In turn, these levels involve issues of how the knowledge mediated in them is applied, of the way they enable self-regulating and self–evolution in us. Through these levels, reflexive considerations of the way systems and drives we are in dispose us to uses of knowledge of us,[16] to cause in ourselves and one another and direction, considerations of issues of what What is us is doing in being us,[17] and the like, emerge.

CHAPTER 3
THE ENCOMPASSING DEFINITION
OF HUMAN BEING

The encompassing definition of human existence that emerges in reflexive human science builds on the minimal definition of human being and the law of human being-process explicated in Chapter 2, that is, builds on the knowledge of 'unitary mechanism'. Expressed in what follows here in several sentences, and explicated further in several paragraphs, the encompassing definition begins by reconstructing the situation of the dialectic of communicative sociation and functional interpellation.

The encompassing definition begins: for subjects structured in unitary mechanism (for experientioning producing-using meaning to make its world the world it prefers as it can make it work), for subjects triadizing in rationalities through narrative-theme-involvement, and that exist in a field of relative experientioning (awareness) in a process producing-using the world-of-understanding (experientioning meaning), human existence is a dialectic of 1) (not always linguistically mediated) communicative sociation (with self and others), of personal- and social-integrations, and 2) interpellation[18] in needs and functional (and other) demands in the system-integrations in the organizational structure of meta-system. Through concern, in the field of relative experientioning, subjects are involved in themselves and other subjects and truth generally, a process involving grounds, necessary and contingent, what is in self-, other-, and object-relations, and that we know currently includes physiology, systems-A/lifeworld,[19] and already added 'systems-B/engineered'[20] which are 'not yet' a perfected survival-system.[21] This is a process involving character and soul, personality and ways, issues of integration (effective coordination), issues of it-with-other and the relevance of aesthetic, issues of structure in added systems-B/engineered and of the impact of these on systems-A/lifeworld (including participant, culture, society),[22] and the rest.

The minimal definition of human existence, explicated in chapter 2 above, in terms of 'unitary mechanism' – the law of unitary mechanism –, the reflexive reconstruction of the definitive objectivity of that, for and in being us, human beings exist as subjects, in will through understanding, producing-using meaning to make the world as preferred as can be made to work, is clearly central in the development of the encompassing definition. Whilst integrating frames of reference such as 'ego' and 'belief/desire-psychology', and 'physicalism', this reconstruction moves beyond them:[23] subjects (instances of unitary mechanism) are there, in narrative-themes, coordinating triadizings (selections and realizations) of value (of preferred ground), there as (generally) embodied spirit (reflexive-subject-in-grounds) in an 'a priori' structured situation of 'systems-A/lifeworld' involving 'a difference'.

Subjects are there in this way in fields of relative experientioning and difference generally, interacting with an environment and participating in structured roles and in some measure adding, reproducing and transforming systems-B/engineered, so, through rationalities, together in a dialectic of communicative sociation and functional interpellation, coordinating conditions of it being one or another way and sufficiently (meeting the demands of) reproducing a paradigm of meta-system(-with-a-difference) which works as a survival-rationality, or failing.

Recognition of the emergence of unitary mechanism (concomitantly of narrative-theme-involved experientioning, in turn concomitantly involving differentiation of subsystems and so on) is the reflexive way, from the perspective of spirit-science, of defining when evolving biology becomes, is definitively, human. (Something like this mechanism may also be operating in other species engaged in the basic fact of life of things counting for or against the viability and well-being of organisms, organisms involved, at a level, in will-to-power[24]). Then, from within the situation of 'biological norms'[25] in which we emerge, communicating subjects manifesting an issue of personal- and social-integration, always already bound in narrative-theme structure, we develop[26] to participant-competence or not. This includes fulfilling roles structured in the organizational structure of meta-system, and – as preferred as can be made to work – adding to and transforming an environment that includes organizational structures and systems-B/engineered, and the patterns of functional interpellation they contain and which manifest issues of integration and drive-logic (including implications for personal and social being[27]), and in which there is or isn't commitment that works, at least, as a condition in the human situation making a survival-strategy (survival-rationality) work. Our relation, as reflexive-subject-in-grounds, to systems-B/engineered, a dimension in 'system-with-a-difference', is a moment in the 'reflexive loop' that we have become in the process of our life: this relation is a dimension in our responsibility and control in our being-process and the becoming, including becoming self-evolving, of our life.

The reflexively aware subject exists as reflexive-subject-in-grounds coordinating triadizings in, and in a relation to, amongst other grounds, physiology, involving characteristics of its being-process (hormonal influences, associated psychological tendencies).[28] Humans, participants, exist experientioning relative to one another as also self-causative absolutes, as individuals, instances of unitary mechanism in-together in an environment involving social needs and opportunities. Reflexive-subject-in-grounds is in a field of relative difference which acts as a source of concern; and participants exist there generating a field of relative value. Evolution acts in the field of relative value, in ways situations of conflict work themselves out, in management of compatibility-issues, in how being-process is occurring, and the rest. The human being-process exists in a process in which a principle of survival-fitness obtains unalterably (is 'a priori'), though criteria of fitness may change. As it is, human existence, a process of the will-to-power-of-freedom, is a property in a life-process in which selection pressures operate on characteristics added, reproduced and transformed through understanding-oriented activity (characteristics including systems-B/engineered, issues of 'how to be', 'what happens', and so on). It is a life in which spirit moves in and between species and their telepathic relations, if not manifests them, and probably exists and moves beyond this as well.[29]

CHAPTER 4
ASPECTS OF PROCESS AND DYNAMICS
IN HUMAN LIFE

At a level, we see that, for us, for the subject, the meaning of being is to make the world as preferred as can be made to work. Reflexivized, we see that this is the case for unitary mechanism aware of itself existing as an adaptive function (with-a-difference) in a species of life-system, an adaptive function (with-a-difference) that is a being-process and that experientions being and applies itself to itself via knowledge of itself and the situation of the life it is what it is in. This, in turn, we see is the case against the background of that surviving is what life is primordially about doing, involving the structured issue of telos in a complete survival-system, along with the dimension in the issue of the subject evolving to exist in ways other than involve biological life or life at all. In the limits of survival and world actually working, the subject's existence in a life-process, a life so being involved, adapted, in production-use of a world-of-understanding, means that successful paradigms of meta-system sufficiently resolve specifiable structured conditions such as meeting all current states of narrative-theme-demands, interpellation-demands generally, and rationality-goals.[30] And it means that structured demands and potentials such as for sufficient preference-harmony can be reconstructed and recognized in the conditions of survival, and recognized in thresholds and extents that vary through time in history with factors such as levels of technological development and mutual dependence, and how rules participants are subject to come into being and operate. Laws are apparent in the conditions of survival, such as a law of increasing demands for preference-harmony with increasing levels of mutual dependence and increasing levels of capacity for mutual destruction, and such as of a tendency to revolution, or at least transformation, with increased repression and opportunity. These laws are apparent in a dynamic in which such as criteria of fitness play a part, along with other variables such as manipulation and control.

For subjects, the environment of narrative-themes and interpellation-demands generally is an environment of participation-conditions: to participate is to be involved in managing, in ways of being, issues (in production-use of meaning), concomitantly, of comprehensibility, truth, compatibility, attitude and orientation, and emotional and spiritual stability and embodiment, and is to fulfil the role of participant-competence in meta-system-process generally. Issues of survival-conditions are reproduced through and select in participation-conditions. It is a challenge and a feedback responsibility, a contingency of survival-rationality in participation-conditions, to recognize and integrate survival-conditions into its ways.[31] The reflexively aware, self-reconstructed subject encounters the limit of value as such, that is, it encounters its situation as it is, as one in which difference, anything, is for it value, 'whether or not also an issue of progress, justice, survival, or whatever'. Because for the subject meaning is (intrinsically) value, fact is concomitantly norm, that is, issue of what it is in world being preferred world. Norms exist at three levels: 1) intrinsic and necessary (for example that a sufficient level of oxygen or integration of compatibility occur), 2) intrinsic and contingent (for example requirement for one or another way of sufficiently managing compatibility or oxygen issues if varied options for doing so exist), and 3) purely contingent.

Ways of being are satisficed:[32] innovated, enthusiastically or reluctantly accepted, mutually established, enforced, and so on. At any moment, participants exist in relation to what is known and practised of 'how to', in relation to issues of 'what there is to know about how to be' and 'what we need and/or want ours to be in', and 'being connected and held in' and 'what should be done', and the like, in general, amongst necessities and contingencies in the conditions and potentials of the process we are what we are in, involving micro-dynamics. So, for example, such as the mechanisms of financial systems are innovated, institutionalised, and in one or another degree regulated; so economic necessities and political choices occur, moralities differentiate and products are created, exchanged, and the rest. Applications of principles reflecting ideology, issues of criteria of survival-rationality, and so on, are involved in the process in which systems-B/engineered (which include language) are innovated, assessed, and regulated, in which situation specifically and generally is innovated, assessed, and regulated: differentiation of a legal-process, democratic political institutions, are examples of ways issues intrinsic in being a field of relative value (involving compatibility and the rest) are addressed and managed. In general, as rationalities are achieved and triadizing is coordinated, the process of the 'ongoing satisficing' of meaning is lived and practised normatively as issue in world being preferred world, issue involving others, truth, institutions, safety, and whatever else. Differences, differences between differences, in general what there is of 'added environment', is there interpellating and providing opportunity for activity through organizational structures. Drive-dynamics are involved in the manifesting and functioning of organizational structures. For participants in them, organizational structures involve issues of relation to conditions of development and personal and social-integration, of relation to institution and reach in their reproduction and transformation, issues of survival-fitness, and so on.

In added systems-B/engineered such as 'legal norms', events, intrinsically, as norms, in involving issues of compatibility and so on, are interpreted and evaluated within frameworks in which they are given particular properties of (culture-linked/ specific) legitimacy/illegitimacy. These properties are delimited in the ways in which, through management − satisficing − of compatibility-issues, the field of relative value is managed as a domain, also, of organized rights: institutions, reflecting and manifesting degrees of such as 'objective solidarity', 'mutual interest', 'domination', and so on, differentiate, also, as places of systematic ways for managing preferred world and its conditions.[33] Legitimacy-mechanisms, institutionalised through the process of satisficing legitimacy, always already involve a relation to the possibility, the ideal, of a mutual unconstrained recognition of a difference, that is, a relation to the possibility that there is mutual unconstrained willingness to participate in an intersubjectively applicable and often binding rule, way, and so on – a norm, in general, for a subject, kind of ground – as appropriate, preferred, as value. This could be called the 'universal value-imperative in grounds'. Short of this intrinsic ultimate potential for complete, universal preference-harmony, the logic in the historical process, at this level of structure, points to a resolution through the differentiation of paradigms within a meta-paradigm. In the meta-paradigm, parameters of mutual tolerance for diversity (amongst paradigms) are agreed and maintained through unconstrained consensus. Short of the ideal of universal resolution, paradigms/meta-paradigm is the most unconstrained organization of world for, the most unconstrained way of, integrating this diversity, that is, it is the least repressive and oppressive, the most stable as far as issues of avoiding the outbreak of violence and war and the like are concerned, because it is the most admitting of preference-harmony and so the most intrinsic within the aspirations for situations-to-be-in of will-to-power-of-freedom.[34]

As it is, ways of being and paradigms of meta-system differentiate and compete in life in a dialectic of domination/resistance-dynamics involving a process of scientific and technological development, and issues of changing criteria of fitness are emerging. As it is human life – contingently – includes domination/resistance-dynamics in the intrinsically structured spectrum (in the satisficing process, in the actuality of doing ways of being) of domination/consensus: the relations amongst participants in the process of the production-use of the world-of-understanding are structured by the spectrum of domination/consensus but can proceed – organize themselves to be – anywhere along this spectrum. In the dialectic of domination and resistance participants encounter issues that include alienation as a result of participation in systems which reproduce the dynamics of mutual destruction (exterminism[35]), issues that include powerlessness in the face of interpellating mechanisms and forces (inability to 'make a difference'); and in it participants encounter the potential in innovating and reaching through the boundaries of conflict-situations with institutionalized mechanisms put into place through socially integrative discourses, mechanisms which are capable of feeding back into and regulating these situations (de-selecting their escalation, enforcing compromise, and so on). This process of the mutual recognition and institutionalization of such mechanisms would set into place systems for managing conflict-situations (for dealing with incompatibilities in the field of relative value manifest through society or however): it is against this background, in this aspect, of an intrinsic potential for management of mutual tolerance and preference-harmony, for manifesting the structure of a logic-dynamic (logamic) that works as the basis of the emergence of a new level of integration, that a need is encountered for making something like the paradigms/meta-paradigm model work.[36] And so, and under pressure from changing criteria of survival-fitness, issues now include a potential for managing the selection of motives in self-activity through new responsibility (managing capabilities in 'being for other reasons' and the like) as opposed to leaving them driven in the drive-dynamics in domination/resistance-dialectic.

Man and technology proceed in the various rationalities that drive, rationalities including, for example, will-to-domination, and profitability. This occurs against the background of the logic in the issue of the survival-system (survival-rationality) being complete, being perfect or not. Value is a limit and a steering device in the situation of the process of meta-system(-with-a-difference); and its reflexive moves are dynamic and determinate: it learns that it questions and develops how things are meaningful, and it can recognize itself operating in relation to, and coordinating, drive-logics and rationality-goals. The logic of perfecting the survival-system always already contextualizes an aspect of the way both the intrinsic, and a current, state of things is meaningful for the subject, that is, contextualizes a dimension of meaning, of the meaning of freedom, of realized value, in history and evolution. So, it evaluates its practise in terms of survival-rationality, and ultimately the principle of perfecting the survival-system. Currently, and arguably in its conditions of survival, the issue of new responsibility the subject has is about its managing itself beyond the supposed determinations of 'a human nature' according to which its ways inevitably manifest domination/resistance-dynamics. This is the issue of a survival-fitness being selected upon by evolution in terms of its capacity for forming, self-evolving, itself beyond domination/resistance-dynamics in a sufficiently holistic consensus-integrated process; and this sufficiently holistic process is possible through, for example, institutionalisation of paradigms/meta-paradigm.

Reflexivized, the subject has an issue of how it inhabits what it is in, and of what it must and can be reflexively related to. For reflexive-subject-in-grounds, a process of 'being scientifically', a process that includes a reflexive therapy, emerges through reflexive human science.[37] The therapy, and a level of reflexive practise generally, emerges through understanding the principle that a person: 'is in, and knowing and orienting (or not) to, the difference a ground makes (or could be making)'. This principle, involving issues in a being-process of what is and isn't there in orientation in 'how to be', in 'range of experientioning', is relevant in analysis in understanding behaviour, and anyway guides in varieties of reflexive narrative: we anyway act on its meaning in ensuring preferred world, in value, as we can make work; we act on it in self-regulation and in facilitating the self-regulation of others in terms of it. A known ground could be 'a necessary condition of survival' or 'how the dreaming works' or 'a tie' or 'the adding or removal of something', or 'a distaste or like for the way a macro-dynamic structure enables a micro-dynamic opportunity', or 'a way of relating to how the life-energies happen', and so on. Aware that it relates to 'the difference grounds make', to what causes in it causing in itself and whatever else, involving such as its reliance on what it learns, memory, and so on, then, the reflexive subject is living the issue it is for itself, is an emergent 'freedom mechanism' being reflexively. It recognizes that, as a structured self-causative self-activity aware that it is emergent at a level in the process of 'the laws of how What-can-be-is-what-is', it is aware that it is relating to 'the determinations of how-What-can-be-happens' in and as it, including in its pure self-determinations, in its existence as 'ground for itself ', and that in its so doing freedom and necessity intertwine. So, the 'idea of freedom' locates its position, positions itself, its necessity and moment, in the encounter with the responsibility that it manage in the mechanistically structured relation to grounds and being caused in, in its situation, potential, and opportunity in the process it is what it is in. And so reflexive-subject-in-grounds recognizes and can develop the role, the use to it in its being-process, of reflexive human science, of this ground; and so it encounters as well that it has issues in how this science can, should and should not, be used.[38]

Becoming the means for managing differences in organized ways, as a feedback system(-with-a-difference), a feedback loop, with integrative powers both in individuals, society, and the systems of meta-system as a whole, reflexive human science reaches the issue of the difference any ground is in the being-process of both individuals and groups, and ultimately in the process of meta-system as a whole. And, so doing, reflexive human science reaches, as well, the issue of that and how the difference any ground is can be managed, an issue involving notions such as 'by letting the market decide', 'through state regulation', 'according to what it means to the individual, and conditioning', 'through social movements', 'in terms of what it means in domination/resistance-dynamics'. So, issues in financial recklessness (both at the level of individual dispositions and system-mechanisms), in violence, in economies adapting within their ecological limits (in 'holistification-adaptation'), in 'new responsibility', in neural networks, in water supply, 'all issues', are integrated. So, along with openness to innovations and the horizons of the not-yet-known, 'maps' of existing and future requirements, potentials and opportunities (conditions of participation and sustainability, for example) can be made and oriented to. So, by integrating conditions of participation, by making it easier to think through the situation of existence and manage mutual recognition, generally to recognize and access and feedback into what a difference means, reflexive human science empowers participants, facilitates, at least through organized understanding and in an objective perspective as far as ideology is concerned, the connection of individuals to control of the grounds of existence, to control of the issues we are affected by, and ultimately the direction of the whole.[39]

Reflexive human science thematizes, in reconstructive and constructive ways, issues relevant to how being-process is occurring: for example, it can be seen that through concern and operating as unitary mechanism being-process has a role in how an organism responds to thirst, it can be predicted that to participate includes being-process in being able to fulfil structured participation-conditions, it can be seen that by knowing about knowing and trying things the subject can produce scientific culture and method and get and use results, it can be seen that how X is being is a ground altering how it is in Y and that Y is managing in it in response, and so on. In general, in the self-application of reflexive-subject-in-grounds, then, reflexive human science is a source of 'being-scientifically': a person is aware of what s/he is in how human life works, and has a reflexive role in how what s/he is works. However, the potential for 'being scientifically' that exists in the process of a reflexively aware subject, potential that can be systematically developed post-reflexivization, and that anyway is a background source of stability, exists for a process, for a reflexively aware subject who, although knowing s/he is a unitary mechanism, nonetheless at moments finds s/he cannot always guide him/herself scientifically in any more than a limited sense. This limit for being-scientifically is realized in terms of, and despite, being self-equipped with reflexive human science: reflexive subjectivity knows the control that can be achieved by manipulating in grounds through a knowledge – in reliable prediction – of existence and process and the difference grounds (including decisions, reach, in sublimation-utilization, in scientific method, and so on) make in it, and can know that at times it cannot know more than that contingency is to be expected.

The reflexive-subject-in-grounds finds, then, that, in being what it is, in being reflexive-subject-in-grounds producing-using meaning to make world preferred world as can be made to work, in fulfilling its role in the maintenance of culture and whatever else, in its own life-chances, it depends as well on such as prudence, on knowing it is working with the unknown and the unexpected (for example the unexpected effects of a new technology, or the behaviour of a commodity in a market, or the result of a sport, or the outcome of an experiment, election, or war, or whatever it is). The reflexive-subject-in-grounds finds, then, that as an aspect of the capacity of being-scientifically, it learns that it is in and depends at moments on such as a feel for it, tact, talent, what happens and happens to come to mind, guessing, instinct, and so on. It finds this because, in aspects of the characteristics of the process and dynamics of its life and situation, it is involved in contingency, including the contingencies of responsibility and creativity and organizational structure, of the unexpected and the limits of knowledge and the enigmas of freedom and necessity.

For us, our story is also one of us making us what we become. Objectively, in intrinsic structures, encountering contingency and organizing and systematizing it, we do to and with one another as we prefer and can make work, and we manage being it in a field of relative value. The issue, in reflexive human science, for reflexive-subject-in-grounds, in the relation of understanding (including desire) and the determination of the will, of what if anything beyond our autonomy determines how we must be if we are to be what we are effectively, the issue of the reasons in there being what we are and how it is what it is, underlies reflection such as Kant's Critique of Practical Reason.[40] In developing the reflexive foundation for human science, here we have reconstructed the law of human being, and, and with awareness' of such as that we can decide to not be at all, and of the role of physiology in grounds, it is in these terms that we address the issue of determination as one of knowledge of structure and causation in the conditions and activities of this being-process. So, we soon find, for example, that unitary mechanism does not operate effectively unless, in being what it is and making culture, it develops rationalities whilst being involved in narrative-themes. And, developing this way of seeing it and doing science of it, and as a basis for understanding the a priori constituents and potentials of a paradigm of meta-system (and including issues of a paradigm running and making its environment/biosphere), and in general as a basis for developing answers to the question of 'how to be it', we use concepts of how unitary mechanism must and can develop rationalities and be narrative-theme-involved, concepts of what rules, if any, we need and can put in place if we are to be able to be what we are effectively, that is, function-with-a-difference and make our world work at all and according to its potentials; we use concepts of laws in the operations of these intrinsic, and intersubjectively interpellating, structures, concepts of necessary grounds, including institutions and other organizational structures, and what these mean in the conditions of rationalities, and so on.

The world can be made less contingent, and much of a level of contingency is removed if the survival-system can be perfected, and is removed as interpellation-effects generally are rationalized and subjects learn about how to be and what is going to happen through how one is being; but if contingency is not an option for people and society, if people and society are not in some measure contingently it, also making and keeping ourselves the people, society and culture, the lifeworld, we are and become, then to what extent can it be truly said that a person or society, the being-processes of unitary mechanisms, the differentiated, made to work, self-integrating, world of narrative-theme-mediated experientioning, human life, is occurring, exists at all?[41]

FOOTNOTES

1   As far as the concept of human science is concerned, in the spirit of the critique
of naturalism, Stan Gooch, for example, [Gooch, Stan. 1975], observes: 'I myself,
together with some other psychologists, consider the wholesale application of
the methods of the physical sciences to the sciences of human behaviour to be
among the major disasters of our time.' In the introduction to his Leviathan
[Hobbes, Thomas. 1968 p83], Thomas Hobbes, for example, observed: (original
spelling is unchanged) 'But let one man read another by his actions never so
perfectly, it serves him onely with his acquaintance, which are but few. He that
is to govern a whole Nation, must read in himself, not this, or that particular
man; but Man-Kind: which though it be hard to do; harder than to learn any
language or science; yet when I shall have set down my own reading orderly,
and perspicuously, the pains left another, will be onely to consider, if he not
also find the same in himself. For this kind of Doctrine admitteth of no other
Demonstration.' With the exception here of the specific goal of reflexive human
science (rather than simply 'a reading'), the reflexive situation is no different
here than that referred to by Hobbes, where the issue of what is generaliseable
in human being is approached through mutually testable (reflexive) knowledge
about being the subject – reflexive experientioning – in it. At a level, with John
Locke [Locke, John. 1977], and later such as Martin Heidegger [Heidegger,
Martin. 1962], and Jurgen Habermas [Habermas, Jurgen. 1984, 1987], this
consisted in inquiry into the understanding. Nietzsche, on the other hand, quoted
in [Hollingdale, R.J. 1977 p 29], influenced by the concept of evolution, refers
to and explicates his view of the family failing of philosophers: 'All philosophers
have the common failing of starting from man as he now is and thinking they can
reach their goal by an analysis of him'. Here, the pursuit of spirit-science, the
distance from naturalism in reflexive science, that is, the process of philosophy
insofar as the medium of observation and experiment is inner ('thought – what
thinks – thinking itself'), nonetheless retains the use of the concept of mechanism
in science. See [Bunge, Mario. 1997 pp. 410-465]. Bunge himself, however,
seeming physicalistically disposed, probably would not recognize the concepts
of 'spirit-mechanism' or 'freedom mechanism'. [Winch, Peter 1958] looks
into developing something like human science via philosophy. Aspects of how
Winch approaches the significance of the concept of rule-governed behaviour
are discussed for example in [Ryan, Alan. 1970 pp. 135-145]. Also [Austin, J.L.
1962], [Berlin, Isaiah. 1997]. Recognition of a field of 'spirit-science' is in the
work of William Dilthey: [Dilthey, W. 1976]. In making this recognition, along the
lines of Locke and Heidegger, Dilthey distinguishes issues of 'causal explanation'
from 'understanding meaning', referring to issues of the latter as issues of

spirit-science (the former being natural science). Also for example [Habermas, Jurgen. 1990], [Plange, Nii − K. 1984].

2   The term 'experiention' (used also in 'experiention-of-being') refers to the existence of being that experiences and intends. Experiention is being that cannot be seen externally but only lived and reflexively recognized during 'experiention-of-being', that is, as we (reflexively) demonstrate to ourselves (in the way we can, that is possible) that and how we know that we are. Experiention is what 'experiences and intends'. What experientions is what has a self, soul, mind, memory, identity-issues, is structured in its being-process, internalizes, and so forth. Experiention is there, us, before, in being a self-relation in it, we begin reflexively to develop knowledge of being it, see it, grounds involved, and so on. Using the term 'experiention' works out a less cumbersome language than that of consciousness, or than language that signifies experience and intention seperately: consciousness is what experiences and intends; there is no experience that is not also an intention and vice versa. That experiention 'relates' − involving a person's being-process − is a problem in symbolic, connectionist, and so on, modelling of the meaning-process in human life. Such modelling, though it is beginning in some degree to manage context-awareness through 'representations of the world' organized in statistics and algorithms, struggles effectively to manage such as the contextual nuance there is in experiential relating, that is, what we live, much of how experientioning is for us. For example: [Wood, Alex L, Merrett, Geoff V, Gunn, Steve R, Al-Hashimi, Bashir M, Shadbolt, Nigel R, Hall, Wendy, 2012]. In some degree, in time, as complexity of response and learning and predictive powers build, it may perhaps be argued that something like a 'value-response' may be emerging in these systems as they 'prefer', 'self-adapt', to dim or brighten etc. according to environmentally sensed inputs: but as far as we know it is not lived experientionally, 'there', there not being a being-process as we live it, are it, in there.

3   The doing of science at all entails that a reconstruction of knowing, and an organization of its potentials, of scientific methods for doing it − involving concepts such as of facts and laws being known, criteria of reproducibility, objectivity, argument, and so on -, occur. Here, method is reflexive, the experientioning that is already involved in science, namely us, organizing an understanding of its existence as such and scientifically through self-experientioning. Oriented in questions/answers such as 'what am I?'/'I am what asks 'what am I?'', it organizes it in a reconstruction of what it sees is the case in an experientioning of itself occurring. Issues of science here are not simply, for example, statistical or concerned with the supposition that knowledge of a 'previous state of matter'

is the basis of a complete picture. (Also footnotes 1, 2 above.) For example, [Harrison, Ross. 1973], [Horwich, Paul. 1990], [Lehrer, Keith. 1990], [Popper, Karl. 1972].

4   The term 'epistemologization-ontologization' depicts the circle involved in experiention-of-being: see experiention, footnote 2 above, and glossary below. In this circle one encounters, through reflexive meaning, the issue of 'knowing of being'. This is the issue of demonstrating the truth in the idea, in the experientioning of the meaning, that one is, in the issue of being, and so of developing 'epistemologized-rationality' to explicate it (also footnote 14 below). Epistemologized-rationality is the knowledge-based process of having and giving reasons. In science, epistemologized-rationality has become a level and mode of orientation in the quest to 'know truth' that is methodologically tied to experiment, involving use of external material technology and observation, measurement, and often, in modernity, mathematical modelling. The issue of the link with truth in 'the production-use of meaning', and so a demand for the demonstration of knowledge, is intrinsic, something soon explicitly clarified in reflexive experientioning (see below, narrative-themes). The explicit clarification of knowledge – a reflexive moment – characterizes a threshold in the process of culture (a dimension of 'Enlightenment'). This moment involves issues of 'realism', here 'admittance realism' (see glossary): clearly what can be known is tied to what is possible for knowing through a method, what can be seen depends on how one is looking, on how the laws of What-can-be-being-what-is 'admit' 'reality-effects' to one's way, and so on.

5   The reconstruction of unitary mechanism is impossible without a generalizable concept of meaning as, in general, What-is. For the experientioner, What-is is 'what-is involving the issue of the 'Whole''. In terms of this concept, understanding can be reconstructed as 'experiention of what-is', that is, as 'experiention of meaning'. Levinas, for example, refers to something along these lines: see [Levinas, E. 1987 p 83]. There is transcendent paradox in the concept of a Whole (It always already is Itself, but is never everything It is, and so on): this paradox is an encounter with a meaning, a boundary for thought, ultimately different from that explicated, for example, in Kant's use of 'antinomy'. There Kant supposedly demonstrates the unreliability of pure reason – involvement in transcendent reference – by showing that that the world began, or never began, can be 'reasoned' either way, so (from the standpoint of reason) undermining the stability in such as the concept of God as the creator of 'everything', of the 'World as Such, as a Whole' (issues of a creator of the world we know nonetheless remaining, along with issues of whether the 'problem' is with 'our' reason or with

the structure of the transcendent referent). See: [Kant, Immanuel. 1965 pp. 328, 297-570]. Here (and more in keeping with the transcendent paradox referrred to above), we suppose the limit at infinity. In general, the reflexive encounter with the 'meaning of meaning' in 'experientionism' (here) is different than that clarified in 'logical empiricism' or 'extentionalism' and so on: see [Mautner, Thomas. 2000]. Quine [Quine, W.V.O. 1953], for example, would probably reject the idea of a Whole as a universal, that is, as questionable inference from beings to Being, to an idea of What-can-be as Such.

6 Through time different aspects have been taken up in different ways by different people. In 'transcendental reflection', Immanuel Kant, for example, examined 'conditions of possibility': [Kant, Immanuel. 1965]. Here, the use, in experiention-of-being, of the relation to there being an issue, is regarded as a more secure, irreducible (bottom line), description of the foundation that there is, in reflexive thought, for certainty of existence, for resolving this concern that occurs in reflexivization, than is possible, for example, with the Cogito of Descartes [Descartes, Rene. 1961], or the absolute of Hegel [Hegel, G.W.F. 1977]. From here, it's reflexive 'self-grounding' [Gadamer, Hans-Georg. 1975. p xxiv] (also footnote 39 below), in the relation of language, thought, and reality, can proceed, as it is possible for it to do, beyond ego etc. (also, for example: [Heidegger, Martin. 1962], [Habermas, Jurgen. 1984, 1987], also footnotes 12, 33), into the mechanics of the reconstruction of the minimal definition, and beyond into the reconstruction of the systems-A/lifeworld in which it, what it is, finds it is emergent in evolution in, and so on ultimately infinitely; and, now living, in-the-world, reflexively knowing itself to be underway, it can proceed constructively further along with structure and potential in its own dynamic.

7 The 'reflexive circle' stems from Heidegger's concept of the 'hermeneutic circle'. From the perspective of epistemologization-ontologization, the lack, in Heidegger, of a clarification of the issue of knowing as concomitant in the issue of 'understanding being' is considered a weakness. Heidegger nonetheless proceeds existentially to reconstruct the 'existentials', the way an experientioning necessarily occurs if and as a world is for it. [Heidegger, Martin. 1962, 1984].

8 The concept of 'being-in-and-for-itself' is in Hegel's 'science of consciousness' as self-consciousness: see [Hegel, G.W.F. 1977].

9 Arguments against reconstructions that use awareness of 'necessary predication' appear, for example, in Nietzsche amongst others, it being asserted that there is no basis for inferring a thinker from a thought. For example: [Descartes, Rene. 1961], [Heidegger, Martin. 1962, 1984], [Nietzsche, Friedrich. 1967, paragraph 484 p 268], [Sartre, Jean-Paul. 1957]. Ultimately, what the issue is – whether it is

of certainty, solipsism, the reliability of a perception, of thinking or doubting or knowing of being or whatever – is irrelevant to what 'that an issue is' means in how the beginning of reflexivization in experiention-of-being works. Physicalism severs the reflexive moment by reducing it. Physicalism assumes that there being a subject that experientions meaning is completely explained by tracing physiological states. Here it is seen that physiological state is accurately regarded only as a ground in there being 'what cannot encounter itself externally, but only be itself ', and in aspects of issues of 'out of body states' perhaps is not always a necessary ground. Whether clarified as such or not, reflexivity is a necessary condition in the development of scientific method at all, i.e. is the knower involved in its issue of organizing its understanding and practise of knowing (whether in a physicalist way or not).

10 The concept of narrative-theme is based on a critical appropriation of, and development from, Jurgen Habermas' concept, in the theory of communicative action, of 'validity-claim': [Habermas, Jurgen. 1984, 1987]. Habermas argues that validity-claiming, as reconstructed in the theory of communicative action, is intrinsic in the conditions of a situation of understanding working at all: such a situation would not be possible if all that was going on in it, in its necessary conditions, was 'strategic action'. Along similar lines, see, for example: [Grice, Paul. 1975]. The difference with Habermas here is not only in that the number of such threads reconstructed is different, but in such, for example, as that 'validating' becomes reformulated as an issue in satisficing (also footnote 32, glossary). For a subject (a unitary mechanism), the intrinsic involvement, concomitantly with the other narrative-themes, in the compatibility narrative-theme, which occurs as it (s/he) is a participant in production-use, is seen to be about compatibility amongst value/power-moments in a field of relative value, rather than, as in Habermas' reconstruction, as always – intrinsically – including an issue of making a validity-claim, claim involving relation to criteria the source of whose authority varies. Here, it is seen that the way conditions-of-participation require of, and provide opportunity and potential for, subjects who share experiention of what-is, means that co-participating in production-use-of-the-world-of-understanding is not simply, or (counterfactually) always also intrinsically, a 'validating' of ways-of-being-it, that is, does not always presuppose that, as a condition of participation, one is 'involved also concomitantly in making a claim, subject to intersubjective recognition, to be being valid' in terms of the 'reciprocated standards of a time' 'that define a paradigm of meta-system', that is, to, for example, be using language 'as it is used, as it is acceptable to so do'. Instead, in general terms, it is seen here that to participate is to be involved in satisficing meaning, which may involve intelligence in making leaps beyond interpreting etc. – operating – in terms of the 'scopes of validity' of existing – established – patterns. Along with at times being demanded of for example to demonstrate why a claim to know something is true, satisficing may also be about 'doing something new, different' without concern as to 'its validity' or 'what others think' (though the recognition – intelligent moves in relation to it, adaptation – of others, generally involving their reference to their standards, and intrinsically whether or not – also – it is found sufficiently compatible, is bound up with its subsequent cultural life and conditions in meaning and being something to them and the process). This way of putting things is here regarded as closer to what is going on, and facilitates discussion of such as the relations between 'validity', legitimacy, and compatibility. That, for a 'demand for validation', there is an issue of a source and legitimacy, has to do with such as established patterns, authority and construction and maintenance of organization. Here, the supposed intrinsic 'validating in satisficing', is reformulated in terms of how compatibility in the field of relative-value works, the management of it being implicated in sufficient levels of integration for functioning at all. Maintaining the concept of satisficing prevents a perhaps rash authoritarian reducing of the conditions of making sufficient – workable – sense together, involving situations such as people with different languages achieving shared horizons, and 'there intrinsically being a validating dimension, implicating a source of criteria' as such. Not anything can work for example in there being civilization, satisficing generates criteria, and these become institutionalized, also, in the protocols of discourses. For a project such as Habermas' reconstructive attempt to generalize moments in 'rational discourse' to the reconstruction of generics in the situation of understanding – of experientioning meaning – at all to be of significance, with what this means in social theory and ultimately human science generally, it is found here that such adjustments are needed: in general terms, to participate is to unavoidably intrinsically be in issues surrounding 'how it is happening' and ''its' being satisficed', that is, 'its' existence (viability, effectiveness, legitimacy, relation to sincerity, compatibility, communicability, validity, and the rest) in doing the narrative-theme-process, or, more broadly, and to anticipate aspects of what follows, in the way a culture of participants is producing-using meaning and reproducing a paradigm or paradigms of the process of a world-of-understanding. In this, there is an issue of generalizeable structure in the satisficing process, a dimension of the issue of necessary structure in any and every paradigm of meta-system. So, here, the reconstruction of what in Habermas is seen in terms of the reconstructive significance of the concept of 'communicative action' (action 'oriented to reaching understanding' as opposed to 'strategic action' – the former though arguable at a level being a kind of shared strategy), occurs in terms of the intrinsic issue of the conditions of sufficient integration through communicative sociation amongst participants in the narrative-theme-process as a condition of the viability of a meta-system-process, that is, in there being a production-use of the world-of-understanding – human life – that works at all. There are moments for example when sincerity in attitude and orientation must occur if the process is to work, as there are moments when participants routinely mutually conduct and refer ('validate') activities according to existing standards. Habermas's perspective has been criticized for neglecting the subject in, for example, [Bowring, Finn. 1996, pp. 77-104]; and obviously here the concepts of unitary mechanism and participant-competence, involving an a priori of narrative-theme-involvement, are central in addressing this balance. (See also footnotes 32, 33 below.) (If Habermas' theory used a concept of 'the reflexive', it would clarify the way terms such as empiricism appear in it.)

11 The concept of unitary mechanism remains apparent throughout. Some religious practice, for example, speaks of letting go of 'human issues' and 'achieving a different level'. Here, from the perspective of reflexive human science with a foundation that includes a concept of unitary mechanism, we see that any move like this is a moment in a value/power-process: the practitioner is always already at a human level, that is, is about achieving a value, a preferred world (in this case 'letting go of human issues and achieving a different level'), if it can be made to work.

12 Similarities with and relevance for aspects of psychology, for example, can be found in evolutionary developmental psychology, in theory of mind concerned with concepts of systems by which desires and beliefs (mental states) are coordinated in actions (in such as looking for things in the wrong place, coordinating action to achieve the value – preferred world – of finding something), in studies of memory, and so on. For example: [Fodor, Jerry. 1987, Wellman, Henry M. 1992, German, Tasmin E. and Wertz, Annie E. 2007]. The case is the same for other disciplines. This science is a way 'the mind' – experientioning – develops thought about itself, answering what is probably an evolved/emergent need the mind has to have a workable theory and practise of itself (taking place here in an evolved state of language, but not limited to it) in terms of which to explain and be itself: the science provides identity with a stability, is a moment in self-reference and mutual recognition, and has the issue of explaining itself in terms of the process of that whose science/practise it is. Issues in this reflexive science are not simply, but can include, statistical issues or issues of the role of a previous state of matter (physiology). Issues, for example, such as of 'the human meanome', arise for this science (meanome being akin to genome at the genetic level but concerned with the 'necessary meaning-structures' involved in being-processes in any and every instance of meta-system, issues arising such as of moments required in levels of development). The term reflexive-subject-in-grounds, along with the concept of unitary mechanism, in some degree represents Heidegger's term being-in-the-world at the level of reconstructing how, in human being, the latter is in the world. As mentioned already in footnote 1 above, examples of projects along similar lines (that is, via philosophy) include, for example: [Winch, Peter. 1958]. Similarly one can say that 'unitary mechanism' is a basis in organizing aspects of the understanding of such as how 'the law of evolution in human history' (a term from the work of Friedrich Engels) works (though Marx/Engels' 'historical materialism' would trace the causes of 'social change' in human history – in the process of meta-system – exclusively to change in modes of production and exchange, that is, in the dynamic of forces and relations of production: see for example [Engels, Friedrich. 2003]). (Karl Marx [Marx, Karl. 2003] also argued that the materialist perspective would ultimately generate 'one science', but not on the lines argued here.) Also, for example: [O'Neill, John. 1981]. In recent times media/communications technologies have made discourse a more significant integrative force in history than used to be possible, altering the conditions of self-evolution (also footnotes 32-36). In general, considerations include such as the relation between the 'laws of nature' involved in the emergence of systems-A/lifeworld, and the way laws in the process of unitary mechanism (involving society) manifest – create, reproduce, transform – systems-B/engineered: (also footnotes 19, 20 below). Unitary mechanism is definitive (of human being), whereas other mechanisms involved in how it functions have varied effects through evolution: for example the evolved readiness to consume once scarce fats has become a source of difficulty in an environment in which now fatty foods are readily available.

13 Discussion of the concept of subsystems relevant here is in, for example: [Parsons, Talcott. 1958]. The term subsystem-with-a-difference is used here because of the notion, in the reconstructive reflexive subject, of a responsibility involving an issue of freedom. Here, responsibility and freedom are seen to be emergent in, and to exist in a relation with, the functioning of systems generally (also footnotes 16 and 19 below), and are recognized through 'unitary mechanism' (the 'freedom mechanism') as properties in the participant-competence subsystem(-with-a-difference). In the subsystems concept, society, for example, is reconstructed as a fixed entity: society is, namely, the definitive gestalt of narrative-theme-mediated activity that the being-process of subjectivity is in, is in in its self-relation and in its intersubjective process, in turn including issues of how narrative-theme-mediated interaction is occurring, that is, issues of culture. To anticipate somewhat aspects of what follows: it is culture that changes, culture that is as it is for subjects in it, who do it, depending on what is possible and occurring, culture whose elements function also if and as polity and economy are, culture that is there through how communication and the rest is possible and occurs. Culture is, is, for example, in the form of a language, if and as the narrative-theme-mediated activity of subjects (that is, society, which need not be involved in using language at every moment) occurs. Culture is what manifests and changes, and changes itself, through satisficing-practise, through the activity, praxis, of subjects linked together in narrative-theme-mediated interaction, that is, who exist in society. Culture is expressed and mediated (becomes interpellated) in technologies, and in other ways. Whether 'personal- and social-integration' (also footnotes 22, 27, below), for example, occurs, depends on whether culture includes ways that are sufficient to the conditions, through 'communicative sociation', of a person and society, depends, that is, on there being the cultural activity/process that occurs in narrative-theme-mediated interaction, in the 'satisficing' of, in participation in, culture. In a sense (and as becomes clearer through the encompassing definition below: chapter 3), as it deploys added systems, culture can make itself impossible, even produce things that can destroy it (below: chapter 4). Personal- and social-integration is linked to how intersubjective and personal communication is possible in how a meta-system-process occurs: in this sense, a meta-system process 'without a society in which personal- and social-integration occur effectively' is one that includes levels of alienation, in which perhaps issues of personal- and social-integration have been reduced to effects of system-integrations, one perhaps that no longer contains domains in which the human properly exists at all, one, for that matter, perhaps which is no longer even 'a meta-system' in 'a world-of-understanding being produced-used'. Culture is if and as society happens, and the properties of culture can vary, with consequences for personal- and social- integration, for what it means to be a participant in it, and for what a participant is participant in. It may be that added systems a culture has become involved in – these added properties of meta-system – happen as grounds that work against the capacity of participants to exist in society or in awareness of 'what we are in this', reducing us beyond these thresholds (exploiting neural plasticity or whatever properties of grounds) to nothing more than appendages in the patterned functioning of things/systems, with relevant consequences. Participation in added systems might also, for example, work against the health of the body depending on such as pollution, the holistic view of meta-system including a sufficient perspective on issues of ecology.

14 Reasons are evaluations lived – satisficed, practised – as norms through narrative-theme-involvement; and one can have reasons for satisficing as one does. A reason, involving a concept of Reason as Such, is a way a subject has of explaining to him or her self and others why something is as it is, involving what this means in preferred world. Such explaining can include explanation of why s/he is doing something, whatever it is, whether it is organizing a logical construction, performing a mathematical calculation, or kicking a ball. In this sense, to be able to explain means for a subject to be able to put to him or her self or another the process of meaning involved, what is being known to be causally efficacious, for example in his or her being a particular conduct. To not know why one is doing something, if it is actually possible at all, is to have lost sight of the reasons for one's actions, resulting in an issue of transparency, a soul searching; and there is always already the issue of whether or not what one is believing is, and/or has evidence is, causally efficacious, actually is what is being causally efficacious. Passion is a reason. And a reason is why a leaf is green or an object falls to the ground (or is levitated) when released. Seeing the concept of reason in these terms avoids confusions and overcomes limits in the concepts of reason that, for example, divide rationalism and empiricism. It overcomes, as well, Hume's, in these terms ultimately confusing, distinction of reason/passion, though without issues for science of such as disinterestedness, objectivity, and so on being lost. It means a concept of 'kinds of rationality' and a concept of 'epistemologized-rationality' emerges (also footnote 4 above). Awareness of that the ways rationalities manifest are shaped by intrinsic structure in the properties of the human situation is apparent, for example, in such as Habermas' sketchy reconstructive concept of 'human interest': [Habermas, Jurgen. 1971]. The concept of an emancipatory cognitive interest, echoes structure reconstructed here in terms of the concept of a 'mechanism pursuing its preferred world', pursuing, that is, its freedom through value-realization. On issues of responsibility and freedom see, for example, Susan Wolf's argument concerning the role of reason in the proper selection of values as the criterion of freedom: [Wolf, Susan. 1990]. On evolutionary epistemology see, for example: [Popper, Karl R. 1984], [Radnitzky, G. and Bartley, W.W. 1993].

15 In Alexander technique, for example, self-application takes the form of an orientation to 'self-use': [Alexander, Frederick Matthias. 1912]. So medicine and other techniques and self-help things past and present, along with the construction and pursuit of opportunity generally, have manifested and emerged.

16 Darwin, Marx and Freud, for example, shifted focus from 'us being us', from concepts of 'bourgeois subjectivity', to include the issue of an account of 'the effect in us being us of what we are in and that is in us'. Systems and drives we are in are a ground making a difference in how we have become and can be and are us, if only as circumstances not of our own choosing (Marx). Issues in uses by us of knowledge of us are thematized, for example, in: Aldous Huxley's Brave New World [Huxley, Aldous. 1988], George Orwell's 1984 [Orwell, George. 2013], and Anthony Burgess's Clockwork Orange [Burgess, Anthony.1995].

More generally, ideas arise of such as genes that won't break the legal-norms, engineering 'unfreedom' and 'not sleeping'. The concept of paradigms/meta-paradigm (see footnote 34 below, glossary) is relevant in the management by us of uses of knowledge of us, involving issues of freedom from outright determination by systems, drive-logics, and evil generally. Paradigms/meta-paradigm emerges in post-modernity as an ingredient in the way through, the self-evolution beyond, the domination/resistance dialectic (the overcoming of exterminism (see footnote 35 below, glossary) and abolition of war generally. Overcoming selection through domination generally involves a new responsibility and addresses an issue of changing criteria of fitness: it is, arguably, a necessary condition in the effective management of sustainability, of economy/ecology in survival-rationality (also footnote 31 below).

17 This structure is reflected in culture in varieties of ways, including, for example, in some religions through beliefs about our being tested in this world and judged accordingly in finding a place in the next, and so on.

18 Interpellation is a term developed from Louis Althusser's use of the term to refer to states others interpellate one in. See for example: [Althusser, Louis. 1972.]

19 The concept of 'lifeworld' emerges in: [Schutz, Alfred. 1980]. This concept is used in Habermas' work. Here, Systems-A/lifeworld is a concept referring to the systems in which human life emerges (ecology, eyes, limbs, and so on), systems which are in the conditions in which subjectivity is possible, exists, and is active participant also in a 'lifeworld'. So these systems are conditions in human life and involve the lifeworld of participant-competence, culture, and society. In the course of the operations of the latter economy and polity are created and reproduced by participants in satisficing-practice (in narrative-theme-mediated experientioning), in turn manifesting the issue of how it – personal and social being and the possibilities for culture generally – is affected by added systems. For us, along with and often linked to the issue of the technical differences they entail, the issue of the effect of added systems and practices is about their consequences in the conditions, health and the rest, of a lifeworld, including our bodies: [For example: Lockwood, David. 1964]. The theme of social- and system-integration is in Habermas (also footnotes 13, 22, 27); the term 'systems-A/lifeworld' is developed in part due to difficulties encountered in the use of Habermas' system/lifeworld distinction as developed by him in the explication of the theory of communicative action: [Habermas, Jurgen. 1984, 1987]. Theories on the origins of consciousness such as Julian Jaynes' [Jaynes, Julian. 2000], for example, is theory tracking the relation, over time, between systems-A and the lifeworld in which our experientioning has emerged and develops and maintains itself, it now being reflexivized.

20 Systems-B/engineered are the systems the life involved in the production-use of the world-of-understanding (us) adds to itself in its process, in its meta-system, as it proceeds, as it can make it work, with making the world it is, and is in, the world it prefers: see glossary. Language is an example of such a system, as are roads, railways, and paints.

21 Life can be seen to be about survival; and, for us, a dimension in the meaning of life, telos in the evolutionary process, can be seen to be about perfecting the survival-system (involving issues such as of whether, and how to demonstrate whether, the supposed laws of 'how-What-can-be-can-be-What-is' being deployed in constructions are eternally the case). Evolution can be seen to be 'the work of something' (involving also 'the laws of chance'), something in 'Reason'; and, though beyond a certain point things become incomprehensible, becoming infinite (inspiring concepts such as 'God the Encompassing'), the issue the reflexive subject encounters amongst its field of concerns, in the structure of value, of, if it is preferred, perfecting the survival-system, remains, a reality and potential for will-to-power-of-freedom.

22 So, for example, paints can pollute, weapons destroy and threaten apocalypse, and so on. So systems can interpellate subjects, so social-integration can be blocked, overtaken by effects of system-integration, so systems (a legal system, for example) are developed to stabilize (routinize, rationalize, pattern) what otherwise would have to be being achieved in communicative sociation, making for new thresholds of organization, and so on.

23 Similarly, Heidegger [Heidegger, Martin. 1962], for example, makes moves into reconstructing the objectivity of the subject that reach beyond the 'cogito' of Descartes and into a perspective on such as 'existentiales'. Though here physiology is recognized amongst grounds, obviously variants of physicalism such as eliminative materialism [Churchland, Paul. 1981] are not upheld: whether or not the concept of a belief or whatever is totally clear to everyone, there is an irreducible issue of 'us being us', of experientioning whose states, including that of there being an issue at all, cannot be measured in any external way but only lived, that is, experientioned and reflexively self-experientioned.

24 Will-to-power is a concept that emerges in Nietzsche: [Nietzsche, Friedrich. 1967]. In this text, put together posthumously, Nietzsche argues somewhat inconsistently, but (like Sartre later: footnote 9 above) there is a thread against generaliseable foundations in human science (hypothesizing the subject to be a fiction: e.g. paragraph 481, p 267), though some interesting ideas relevant in the foundations of human science emerge.

25 On 'biological norms' see, for example: [Eibl-Eibesfeldt, Irenaus. 1979]. Concepts relevant there include such as animal intelligence.

26 Developmental theorists such as Piaget (pre-operational, concrete, etc.), and Kohlberg (pre-conventional, conventional, post-conventional, etc.), examine developmental stages: here stages are thought of as stages in the process of the emergence of the subject as participant-competence, that is, as capable of its being-process, involving its role in meta-system-process. Developmental psychology is relevant here in understanding the process of development in the coordination of attitude and orientation (in the subject), in stability in embodiment, awareness of truth, capacity to be comprehensible, and the like. There is evidence that a human infant left to develop with dogs will become irretrievably dog-like, subjectivity as we know and are it failing to manifest, lending weight to, for example, Locke's 'blank-slate' hypothesis. This suggests our existence is a structured potential, a necessary pattern (mechanism), for experiention in the appropriate environment of systems-A/lifeworld; and it suggests that a socio-cultural dynamic, stemming from a level of animal intelligence/biological norms, is a condition in its occurring (involving individuations, self-identities, intersubjectivity as ground, constructions in managing social needs). Seen as elements of a 'transcendent normativity', such as the reconstruction of narrative-theme-structure reconstructs the 'a priori' structure in the potential becoming and being actual, there being socio-cultural conditions in the conditions – grounds – of the phenomenon. All this is consistent with the organization of human science in a reflexive spirit-science, as argued here: involving brains and whatever else, spirit –an experientioning being-process with responsibility in itself for itself – is seen to self-differentiate and absolutizes in the self-relation in self-presence; and in this identity, for example, involving the way internalized environmental-factors are managed, is key in how evolution operates selection-pressures through motivation and in activity generally, this generally being something in how how-What-can-be-can-be is a life-process adapted in world-experientioning with potential for reflexive self-application and self-evolution. In evolved organisms, the production and reproduction of innovations, maintenance and self-transformation, is a process in subjects (of a spirit-mechanism structuring experiention), subjects somehow finding connections (perceptions, linked-presences, and so forth) through a spirit-medium ('world soul'). Spirit – the substance that experientions, which is not apparent 'externally' but which is affected in grounds and that is there only as a being-process, in and for itself – perhaps exists also 'outside of – across – space-time', and to some extent is itself active in the ground (for it) of human life (and life generally). (This perhaps prompted aspects of thought in such as, for example: [Standing, Herbert, F. 1930]. Standing (p13) comments: '..the whole evolutionary process is fundamentally a manifestation of Divine purpose and activity..'. Though, in Standing, the term Divine is unclear and contingent, here the concept of 'something-about-how-What-can-be-is-being-what-is, despite 'transcendent paradox' (see footnote 5, above), is not. It is clear that somehow spirit and matter meet and interact in the human (and other) organisms, and do so in the conditions of experierion at all. And this – involving such as 'the thating of 'the that'' – is consistent with what some anatomical features of the brain are probably for. Spirit-oriented strategies, for example interference in the operations of the subject (in how 'how being-process is' is), as well as more directly anatomical moments, are relevant in evolutionary process.

27 The integration of personal and social processes, involving such as management of sublimation-utilization orientations and thresholds, is bound up with the demands of participation in the environment of systems-B/engineered, which in turn manifest issues for survival and ecology. In coordinating its being-process, which is structured and cannot happen anyhow, the subject faces a variety of issues and at a variety of levels.

28 Issues relating body-chemistry and motivational tendencies are widely discussed, including, for example, in a BBC2 television programme following the credit crunch of 2009 called 'The City'.

29 The theory of evolution according to chance mutation and natural selection asserts that, if the time frame is long enough, then it must be that, because chance mutations are plentiful enough, self-replicating genetic material does what it does, and is selected for or against doing it according to the fitness, in environmental conditions, of the characteristics expressed in the organisms involved; and the theory argues that this accounts completely, for example, for the eye starting out as it did and becoming what it has, or for the emergence of wings and hearts. See: [Darwin, Charles. 1996], [Dawkins, Richard. 1976]. It can seem unlikely that this accounts completely for the properties and directions in life we know of, and it leaves unaddressed the fact of the emergence of life at all: what is doing this are the laws of how What-can-be happens at the level of the life-process, also perhaps involving chance, that is, now, the systems-A in which, for example, the lifeworld we are in exists. There are levels of how it is happening, involving us sometimes in encounters with such as 'manipulated coincidence'; and it is What-can-be-being-what-is, involving perhaps the fulfilment of 'to be' (of the meaning of Being) in a completed survival-system. Now, through knowledge of genes etc., reflexive-subjects-in-grounds (we) are increasingly responsible for direction. Because, at a level, reflexive-subject-in-grounds is a differentiation of it (It) within its (Its) whole (Earth), the whole (and ultimately in the form of it being made a perfected survival-system) is increasingly an issue for the former, now 'absolute', and also self-evolving director.

30 So, for example, the system of language (an added systems-B/engineered) must work in the experientioning of what-is well enough at least for survival if the life-system adapting to use it is to survive with it included amongst its strategies.

31 So currently, for example, issue of the sources of global warming in ways systems-B/engineered happen, prompts adaptation in participation-conditions (in what it means to be doing, and so in kind of system-B/engineered we are in) in the interest of harmonizing them with survival-conditions, involving the issue generally, in survival-rationality, of the management of the relation of economy to ecological limits. Knowledge useful in learning of ecological parameters includes such, for example, as: [Marshall, Michael. 2011]. See also glossary: holistification-adaptation.

32 To satisfice (also footnote 10 above, glossary) is to be a subject involved in working out (including mutually establishing) the meaning it is involved in (in which an intrinsic issue of value is encountered), and institutionalizing what norms it does (i.e. the process of ground-relation, which includes language-usages) in which its being-process organizes its existence. Satisficing is concomitant with triadizing in the process of unitary mechanism. Satisficing – a process of self-involvement and intersubjectivity – happens in every encounter with meaning, that is, is more or less continuous, what happens as one – experientioning – is, and involves the five narrative-themes concomitantly: 'Being' always already is satisficing if and as a human being-process is occurring. Culture is the outcome of – and is reproduced through, exists as – 'a satisficing-process'. Satisficing as used here derives from the term in psychology coined, according to the Penguin dictionary of psychology, by Herb Simon; there referring to an acceptance of a judgement or choice as one that is good enough when optimal solutions are not to be had. Here, it refers somewhat conversely to that for us to be is to be 'making the meaning as one prefers and can make work', though a satisfice, a triadized value, could be, but need not always be and often isn't, an 'optimal solution'. Often, one has to make do with other than one's preferred ground as such, the value triadized being to do the best one can under the circumstances, that one can make work. In developing this term, one can move, in the foundations of the reconstruction of what it is to participate in experientioning meaning, involving concepts of fact/norm/value and so on, beyond the comparatively restrictive concept of 'validity' and so 'a limit at validity-claiming', and into a broader concept of the necessary condition that is participation in narrative-themes, which can include such as mutual recognition of legitimacy, tradition and an organizing of 'valid', and so on. Because narrative-themes are ongoing, and attitude in participating in them can be transformative of or transgress existing forms, the concept of always already intrinsically being in a demand to be 'claiming' 'to being valid in how one is being' 'if one is' is not sufficient: the latter can be a part of a project like 'the construction of a notion of communicative action' [Habermas, Jurgen. 1984, 1987], but at the foundations of a reflexive reconstructive human science as such, are not sufficiently clear and 'true to life'. For the general terms of this science, a concept of satisficing in narrative-theme-involvement, involving the reflexive dimension of ones relation to oneself (and others) in it, is here found to be a required generic, a more adequate generalizable description of how participating (in being oneself and with others) – being-process – is. The world can work whether or not a satisficing of attitudes such as sincerity amongst participants in it involves a thematization of the validity of a claim about it.

33 Jurgen Habermas [Habermas, Jurgen. 1998], for example, examines aspects of the differentiation of a legal process. Habermas' discussion roots itself in the concept of the separation of facticity and validity in the process of experientioning (meaning): on this separation as conceived, before Habermas' theory of communicative action, at the level of 'normative system' and 'patterns of behaviour which people actually perform', see for example Schneider, David, quoted in [Douglas, Mary. 1975 pp. 205-6]. Here, (also see criticism such as that of Finn Bowring, footnote 10, above), the discussion of this separation, and of the linkages of facticity and validity in natural language-use traced by Habermas in the process of thematizing validity-claims, is considered a way of discussing an aspect of what is involved in satisficing (also footnotes 10, 32, above, and glossary). It is discussion which, here, can be thought in terms of the reconstruction of the subject as 'unitary mechanism', and of compatibility as a narrative-theme (a narrative-theme happening concomitantly along with the other narrative-themes in terms of which world is encountered and value is organized and concern – what is pressing etc. – resolved). Here, the subject (and its collective groupings) is conceived of as an experientioning entity for which – as a spirit-mechanism in which – fact is also always already norm, that is, as an experientioning entity in which (experientioned) meaning is always already issue of being ground in preferred world, issue of value. The fact, in human life, of 'experientioned meaning as also always already value' (that is, issue in preferred world being made to work), is a concept that involves an intrinsic sense of 'normative' in the concept of meaning at all. For the subject meaning at all is normative; normativity is conceived as intrinsic in experientioning, the latter involving in value-orientation in a more general way than simply, for example, as is apparent in the notion of justifying a source of ideas of right and wrong, and of rules and regulations actors should arrive at and recognize together and follow in pursuing self-interests. The critical appropriation and development of Habermas' theory of validity-claiming in the foundations of reflexive human science in terms of the concepts of narrative-themes and satisficing practice (as outlined above in footnote 10, i.e. also via a reconstruction of the subject), indicates that the theory of the subject as a unitary mechanism linked to the world and coordinating its value through the narrative-theme-mediated encounter with meaning, is presupposed in the former. The theory at the reflexive foundations of human science is a reconstruction that, through the concept of 'integration through the narrative-theme-mediated interaction of subjects (of unitary mechanisms)', makes clearer both the structure of the situation itself, and the context in which a hunch about an immanence of 'validating' as an intrinsic orientation structuring the conditions of communicative action being, (if only via a background consensus), a condition of possible unconstrained understanding at all, a generic in the structure of how human life works, makes sense: it is because we are unitary mechanisms intrinsically involved in narrative-theme-mediated interaction, and with a capacity for reflexivity and examination of how what we are is possible and works, that we can construct/reconstruct notions of 'an ideal type of communication' based on a reconstruction of how utterances involve an issue of validity and ideas of 'the conditions of a possible understanding', that we get involved in issues of the mutual validity/legitimacy of a value in the field of relative value, etc. The concept, here, of norm and meaning, in terms of which meaning is concomitantly norm because it is intrinsically value, ground in the preferred world of the narrative-theme-mediated interaction of unitary mechanisms, then, reaches through (and encompasses) the concept, in relation to nature, in society, and in the inner life of individuals, of the validity of utterance through validity-claiming (validating). It does this in a way that, reflexively, we need to see and be aware of and practice: for the subject, any meaning, which anyway has the objective relation ('facticity') it does in the way world generally is working, always already also works as a norm, that is, exists, works, as a ground in the continuum of the issue of preferred world. Because a meaning works like this, it intrinsically invokes an attitude of value (in the subject) towards it, and intrinsically is something in a field of relative value (a normative issue – issue of compatibility – for others). A subject – a participating competence – is a unitary mechanism (linked to the world through narrative-themes etc.), whether or not s/he is involved at a particular time in identifying facts, or in sorting relative values in terms of an ideal of validity and an imperative – a managing – of integration, or in dealing with situations of conflict by regulating 'strategic activity' by establishing and mutually recognizing norms through the protocols of 'communicative action', or in internalizing identity, or whatever it is. Looking, in the same perspective, at a particular angle on fact/norm, one sees as well such as that the situation of perceiving a fact at all generally also involves organizations of value, that is, organizations of ways being-process can make meaning work: norms such as reproducibility, disinterestedness, falsifiability, etc. in terms of which experiention is organized – organizes itself – in perceiving facts are apparent, for example, in scientific method: [Popper, Karl. 1972]. For reconstructive reflexive human science, then, the concept of communicative action addresses issues of what transpires intrinsically in interaction through natural language-use. Read as project that transposes/ generalizes the mechanisms of (the validity-claiming reconstructed in) the process of academic research ('rational discourse') into the project of reconstructing the conditions of human activity generally, much of Habermas' project is perhaps to be significantly interpreted as constructive, as intending to steer a direction, via a reconstructive understanding of the significance of discourse-pragmatics in human life (universal pragmatics), 'toward a rational society' [Habermas, Jurgen. 1970]. Such a 'rational society', a morality of discourse and consensus, is one in which it is thought that differentiated and pluralized lifeworlds can be sufficiently socially integrated if legitimacy is established and intersubjectively maintained, and 'strategic activity' is regulated, in rules arrived at through communicative action, in discursive redemption of validity-claims (i.e. through unconstrained mutual recognition rather than in rules sourced in other ways, often arbitrary and involving domination through force and attendant legitimation crises).

Here, then, for the purposes of the foundations of reflexive human science, Habermas' universal pragmatic approach to the reconstruction of the conditions of possible understanding, is recognized to be a very useful aspect to develop and critically appropriate, but ultimately in itself to lack sufficiently balanced emphasis on the subject in favour of its intersubjectivity. A reconstruction 'of how a level of human life is possible and works' (of the mechanisms of a level of human life) undertaken 'beyond philosophy of consciousness' using the theory of communicative action and the system/social distinction in integration, Habermas' account, through several texts, of universal pragmatics, is an account of an intersubjectivity of mutual recognition working, for self-interested actors who also evaluate facts in the light of their preferences [Habermas, Jurgen. 1998. pp 26-7], potentially as a dynamic of communicative action in which utterances are validated: as such it reads also as an argument concerning the conditions without which a socially integrative shared understanding would be impossible, and as an account of the kind of communication (the kind of assent and so on, understanding) that would be involved if we could be sure that we were achieving an unconstrained relation (see also footnote 34 below). Developed here in the reflexive reconstructive project (see also footnote 10 above), Habermas' is an account that can sometimes seem to be of the same thing (a subject involved in facts and in norms) as though it were two different things, for example: [Habermas, Jurgen. 1998. pp 26-7]. No such conflation or division or lack of emphasis on the subject as such, emerges in language that includes both the theory of unitary mechanism and of narrative-themes, theory, here, that includes the reflexive discussion of preference and value (and relative value) in experiention-of-being and the theory of unitary mechanism, in conjunction with the reconstruction of the meta-context of narrative-themes, satisficing, etc., and which is theory that is required in critically appropriating anything like the theory of communicative action in reconstructive terms in the foundations of reflexive human science.

Subjects are involved in being 'how it is possible to be what a subject is', and this entails intrinsic objective structures: a person does as s/he does, and what s/he does is what it is in and for his/her life and the lives of others. The differentiation, in the evolutionary and historical process, of a subjectivity, and of the operations of its narrative-theme-involvement, is concomitant with and an emergence within a social life proceeding previously only in 'biological norms' (see also footnote 25 above). The capacity to reconstruct such as a legal-system and its role and how it works in a meta-system ('societal') process, is not lost in the theory of unitary mechanism and meta-system(-with-a-difference): this theory links polity with the differentiation, in the polity subsystem, of ways of managing—doing—compatibility issues, such as a legal-system, compatibility issues which exist intrinsically in the field of relative value. Managing compatibility-issues is a process that includes issues and attitudes of 'validity', 'legitimacy', 'domination', 'consensus', and the like. This theory is aware of, and clear on, that both 'fact' and 'norm' are value for a unitary mechanism, and are intrinsically linked in the conditions of culture generally (at all levels), as well as to motives and an ideal of preference-harmony. It can easily be seen, in terms of this theory, that, in the differentiation and development of institutions, issues of drive-logic and direction manifest through system-imperatives such as profit; and that, as a dimension in issues surrounding the administration of ways for thematizing and managing compatibility issues (in satisficing value), the process of a legal-system is no exception.

34 We each are, humans are, instances of unitary mechanism, and exist in relations together in a field of relative value: within the constraints of the issue of the conditions of sufficiently effective satisficing that a world can be made to work at all (including dimensions of coordinating 'mutual validity' and so on), it is up to us how we manage contingency in the field of relative value. Jurgen Habermas [Habermas, Jurgen. 1984, 1987], who reconstructs an infrastructure of 'validity-claiming' as 'a priori' in satisficing, in how narrative-theme-mediated experientioning works (see also footnotes 10 and 33), can be read, also, as an examination, in terms of the issue of the kind of communication (the kind of assent and so on, understanding) that would be involved if we could be sure that we were achieving this universal unconstrained relation, of the ideal of a mutual unconstrained recognition of a difference, can be read, that is, as an examination, in this way, of the 'universal value-imperative in grounds' in terms of the possibility that there can, or must sufficiently, be mutual unconstrained willingness to participate in an intersubjectively applicable and often binding rule, way, and so on – a norm, in general kind of ground – as appropriate, preferred, as value. And this, somewhat ironically, could be useful in developing and institutionalizing a 'communication-system' whose protocols interpellated participants as would a legal-system (see also footnote 36 below). Paradigms/meta-paradigm is a model, implying a directional logic of development – a telos – in the conditions of freedom toward increasing preference-harmony. This way of managing the field of relative value, this possible institutional form for managing in the field of relative value, is probably required in practise in the abolition of war. In paradigms/meta-paradigm, paradigms, versions of meta-system, exist divergently if preferred, and so exist without interference from each other. They exist this way because, as participants in a meta-paradigm in which parameters of mutual tolerance for diversity have been agreed, they are freed from pressures of selection upon one another through domination. The institutionalization of a meta-paradigm is an institutionalization in which selection through domination has been de-selected: instead, what paradigms prevail do so through the mutual recognition of norms that have been discursively redeemed in unconstrained consensus concerning parameters of mutual tolerance, probably involving compromise. This means that, when there is diversity of value about conditions of participation, then being-process, world, participant-competence, the subject, is freed through integration in paradigms/meta-paradigm. It is so freed 'post-modernistically', that is, it is freed beyond any meta-narrative other than the meta-paradigm 'generalizing' itself, its justice etc. Though Lyotard questions the notion of consensus from a perspective concerned with what it could mean for content, for post-modernism see: [Lyotard, Jean-Francois. 1987]. In paradigms/meta-paradigm, it is freed a stage further than is possible in the modernist situation of a force that generalizes a paradigm of meta-system without the possibility of the differentiation, within a meta-paradigm, of other kinds of paradigm, i.e. despite that a preference for 'other ways' might be the case within some of its participants; and a meta-paradigm-process checks as far as the content of a consensus is concerned by integrating mutual interest in mutual tolerance and an opportunity for diversity, meaning that consensus can integrate diversity, and that there is always an option and the pressure of selection through a criterion of mutual tolerance.

Compromise limits options, and it is likely compromise will have to be a mechanism in making paradigms/meta-paradigm work, should it occur. The mechanism of consensus-integration in the construction of paradigms, ensures that participants in a paradigm want it as it is, are determining their grounds for themselves as they can make work, and are free from being selected on through domination by others in what they are in. Paradigms form as participants bind through the mechanism of consensus-integration; and, concomitantly, paradigms integrate through consensus concerning parameters of mutual tolerance in the forming of the meta-paradigm. Rather than people being bound by a generalizing model anchored in a ratio of consent and domination, differentiations of diverging paradigms integrated within a parameter of mutual tolerance are possible. So, for example, as discussed in footnote 33, just as Habermas ties legitimacy to the concept of the discursive redemption of validity-claims, a paradigm exists within itself as a consensus-integration along these lines (according to the freely integrated value of participants in it), and paradigms exist similarly within the meta-paradigm. Without a concrete integrative concept and praxis of paradigms/meta-paradigm, such as Habermas' perspective has little actual potential in the process of the unfolding of the will-to-power-of-freedom, in history and evolution. As it is, the more systems of domination are needed, and the more warfare is involved or at least (via the arms race) a potential (which nonetheless from some perspectives might be desirable), the more repression/oppression generally is indicated in the relation of a participant's value and the actual conditions of existence in the dynamics of the domination/resistance-dialectic: the maximization and stabilization of preference-harmony in paradigms/meta-paradigm is the alternative to, the way out of, a drive-logic driven trajectory to apocalypse and, potentially, that includes an Orwellian nightmare, or the unintegrated instabilities of rival groups and factions. According to such as Fukuyama – [Fukuyama, Francis. 1992] –, for example, every paradigm in a meta-paradigm would probably choose to be a democracy of some kind. Also, for example [Tilley, Christopher. Y. Miller, Daniel. Rowlands, M.T. 1995], [Rees, Martin, J. 2003], indicates a need for a sufficient global consensus-integrated stability.

35 Exterminism refers to the process of an arms race, involving the production of ever increasing capacities for mutual destruction and total annihilation, and concomitant demands therefore on economic activity and strain on ecological integration. It is a term developed by E.P. Thompson: [Thompson, E.P. 1982]. Exterminism exists because of how contingency in the field of relative value is managed and structured. It expresses the drive-logic of domination/resistance-dynamics: about 'more to be better at war', it manifests issues for survival of overcoming domination/resistance-dialectic, issues of meeting the challenge of changed criteria of fitness through sufficient consensus-integration and new responsibility. Arguments exist which link it – along with aspects of the phenomena of war generally – to such as disaccumulation mechanisms in the overproduction-crises of capitalist systems (state-managed demand). For example: [Harvey, David. 1982], [Rousseas, Stephen. 1982]. As it is, increasing efficiency of systems is creating the situation in which not as many people need to work in the systems, as the functioning of the systems is able to support, putting pressure on existing organizational structures. As it is, and with further greening of technology, over-production crisis only indicates an increasing potential for freedom from war and work through increased efficiency. If war is abolished, which could be coordinated for example beginning with a global referendum, and through stability in meta-paradigm/paradigms, actualizing the potential in the issues of freedom and work is a matter only of value in what is possible and can made to work in terms of these new, increasingly efficient, technologies, and now without the restraints of an exterminist environment.

36 To conceive that there is no social/system difference, as for example does Mouzelis [Mouzelis, Nicos. 1997. pp 111-119], is here considered an error. Accepting, in the reconstruction of human existence, the theory of it as a meta-system-with-a-difference involved in the production-use of a world-of-understanding, and in that the theory of unitary mechanisms involved in narrative-theme-mediated experientioning and rationality-goals, the difference between social-integration and system-integration is clear. The difference is encountered regularly each day by more or less everyone: one encounters a system-interpellated moment, for example, in the demand for a money-exchange as a condition in purchasing food at a supermarket (system-integration), though a casual conversation does not have these properties (social-integration). A lighting-system is a system one uses, and, for as long as we are involved in the decision, in deciding whether or not to switch on a light or to save electricity, or whatever it is, involving stabilities, one is coordinating oneself personally and socially, with-a-difference, in relation to it. In paradigms/meta-paradigm, the development and institutionalization of a system of rules for the regulation of activity/conflict, would, in the creating of and participating in the discourse involved, have to involve social-integration, if only in developing and institutionalizing a 'communication-system' whose protocols interpellated participants as would a legal-system. Getting to a 'world beyond peace and war', achieving the abolition of war, could, for example, involve a 'communication system' put into place in this way. Put into place, along the lines of Habermas/Apel's 'unlimited communication community' [Habermas, Jurgen. 1984, 1987], [Apel, Karl-Otto. 1987], as an institutional basis for satisficing mechanisms for managing conflict situations, for ensuring compromise and so on, arguably this evolution, the type and scale of communication-process and consensus-integration required for sufficiently getting out of domination/ resistance-dialectic, is increasingly a condition in managing ecological-integration and survival-rationality generally, and from this perspective should be a moment in participation-conditions. Realizing the other moments in coordinating survival-rationality, in turn could involve such as an economy of carbon-credits, a quantity of such credits being allocated to each individual equally as a birth-right (human right), and the like.

As far as 'complex systems theory, for example' [Mitchell, Melanie. and Newman, Mark. 2002], is concerned, and against its background distinguishing 'complicated systems' (like a car) from 'complex systems' (which aren't quite 'systems' in the sense of a car, the functions in which are, at a level, all always predictable) it may be that the properties of 'system-with-a-difference', of subjectivity (unitary mechanism), ultimately mean that the meta-system-with-a-difference of the process of life producing-using a world-of-understanding cannot be sufficiently effectively modelled in mathematical terms, meaning that only the reflexive foundation for human science can integrate the explanandum of human life, knowledge of the regularities, 'emergent behaviours', in meta-system-with-a-difference that are known through complex systems theory so aligning in and for reflexive human science, like anything else, as a knowledge of grounds for reflexive-subject-in-grounds, which, for example, so knows about how to feedback through the controls of the reflexive loop in order, for example, to regulate in systems-B/engineered to minimize the pathways for epidemics (i.e. which so is better placed in rationalities involved in realizing a value), and so on.

37 See: [Mortimer, Chris. 2013].

38 Issues in the uses of reflexive human science are much as issues in the uses of any science (see also footnote 16 above).

39 In the post-modern era, in the management of integration 'post-meta-narrativity' in the running of human life, there is an issue of a need for a method for mutual recognition emerging along with the potential for pluralized lifeworlds, differentiated ideologies and paradigms, initiatives 'beyond peace and war', and so on, a method in which reflexive human science is relevant. This is because, in the self-understanding of the post-modern era, notions of universalized, generally applicable, 'validity' and 'legitimacy/illegitimacy' and the like become eroded in favour of, or contextualized in, notions of 'things operating and emerging if and as a value/power is making – satisficing (using discourse, domination, consensus, etc.) – it so, is becoming a result of its own determinations, involving reflexivities [Lawson, Hilary. 1985] and compatibility issues, and is making it so for reasons that include such as experience, control, system-imperatives and –potentials, the rationality of drive-logics and systems-of-domination, learning' [on learning through experience, for example: Hume, David. 1988], meaning that something like meta-paradigm/paradigms is required in the possibility of the integration of potentially varied post-modern ideologies and paradigms, and in turn meaning that, along with new responsibility, something like reflexive human science is required if pluralized lifeworlds are to be able to mutually coordinate their respective sensibilities and identities whilst sufficiently integrating potentially varied ideologies and paradigms.

This, despite the obvious meta-aspect to the concept of meta-paradigm and the generics of reflexive human science, a meta-aspect that reaches beyond, is not the same as, the 'meta-narrativity' criticized in post-modern critique. Removing the compulsions of selection through domination, meta-paradigm/paradigms both ensures and stabilizes the possibility of diversity of ideology and technology in the 'post-meta-narrativity' dawning, checking potentially self-destructive tendencies in the whole and preference-harmony by integrating mutual interest in mutual tolerance and an opportunity for diversity. The survival-need is one of ensuring stability, unity in diversity (assuming a diversity of paradigms), in the field of relative value, sustainability, beyond such as motives, rationalities, driving exterminism (the exterminist meta-narrative) etc. Meaning that consensus can integrate diversity, that there is always an option and the pressure of selection through a criterion of mutual tolerance, meta-paradigm/paradigms prevents an enforced hegemony putting an end to ideological diversity (should it be value), addresses the conditions of the issue of the inevitable emergence of a technocratic domination as a result of involvement in science at all, or the domination of the technocracy of one or another science and technology. And in some degree, as a basis for mutual recognition amongst pluralized lifeworlds, reflexive human science underpins the possibility of integration occurring in anything like meta-paradigm/paradigms. For example: [Bell, Daniel. 2000], [Farganis, James. and Rousseas, Stephen W. 1963], [Fukuyama, Francis. 1992], [Gouldner, Alvin W. 1976], [Lyotard, Jean-Francois. 1987], [Ravetz, Jerome R. 1971], [Rees, Martin, J. 2003], [Thompson, E.P. 1982].

40 Using concepts such as the categorical imperative and the moral law, Immanuel Kant [Kant, Immanuel. 1996 pp. 32, 46, 49, 92, etc.], for example, reflects, in what can be recognized in some degree as a reflexive and reconstructive way, on the notions of self-causation and how we must act, and in this, distinguishing morality and legality, on the issue of a transcendent source for issues of conscience and feelings we get in us depending on how we act, and so on.

41 In a hermeneutical philosophy of the human sciences, Hans-Georg Gadamer, for example, [Gadamer, Hans-Georg. 1975], discusses such as Bildung. He can be read to project the issue of the relevance, development, and maintenance, in human science, of a paradigm based on the inner self-activity and reflexive self-grounding of a process of being (a being-process, experientioning) that understands and that involves language and an infinite relation, and that reflexively develops and maintains, integrates, its self-relation and its internal connections and communications with others, in its society and culture, its history and world. The reflexive approach developed here is opposed to a notion of exclusive reliance, in human science, on the methods of a science mobilized technocratically exclusively through sense experience – at an 'empirical' distance, physicalistically, non-reflexively – in an examination of 'cognition' that quite readily forgets, if not is methodologically bound to systematically exclude, being and value. Here, in a reflexive foundation for human science, then, it is clear that and how reflexive-subject-in-grounds is a self-experientioning being-process, and, in clarification of the relation to physiology or whatever it is of grounds, it is clear that it can include physicalist knowledge or whatever knowledge of what it is that is in it. Here, it is clear that reflexive-subject-in-grounds can deploy physicalist knowledge or whatever it is with understanding of the difference doing so is to it, in its being-process, that is, without losing hold, concomitantly, of that it can only be itself and of that it encounters contingency in this, of that it encounters the difference its self-causing makes in it along with the issue of what if anything else might be causing in its experiention of being self-causing. So it is apparent, for example, that 'cognitive bias' can operate in how we think, and that reflexive-subject-in-grounds can learn of and test for this about its thinking and look out for it and so on, can see of a ground amongst grounds that could be functioning and relevant in such as how rationalities form, in what is in a reason: [Kahneman, Daniel. 2012]. Once aware of such ground-effects causing in how it works, the reflexive-subject-in-grounds can both reach beyond it through value/power and rationalities, for example by deliberately paying more to disprove that 'loss aversion' is a determination in its rationalities, and, to the extent it is possible, it can 'treat' it with techniques of reflexive being. So, reflexive-subject-in-grounds can utilize physicalist knowledge or whatever it is without losing hold, anyway, of that it cannot locate itself as such anywhere externally, for example by looking for itself with a microscope, but only reflexively, in self-experientioning and being itself, including in self-applications in which physicalist knowledge or whatever it is is learned of and used by it, and mediates a difference in it, enabling it, for example, to know more about and self-apply in, be scientifically in relation to, what is going on when it experientions that it moves a limb, thinks, causes itself to be, or whatever it is.

REFERENCES

Alexander, Frederick Matthias. Conscious Control, (London, 1912).

Apel, Karl-Otto. "The Problem of Philosophical Foundations in the Light of a Transcendental Pragmatics of Language", in eds., Kenneth Baynes, James Bohman, Thomas McCarthy, After Philosophy, (The MIT Press, Boston, 1987. pp. 245-290).

Althusser, Louis. "Ideology and Ideological State Apparatuses", trans., Ben Brewster, in, Lenin and Philosophy and other Essays, (Monthly Review Press, London. 1972).

Austin J.L. How to do Things with Words, (Clarendon Press, Oxford, 1962).

Bell, Daniel. The End of Ideology, (Harvard University Press, 2000).

Berlin, Isaiah. The Proper Study of Mankind, (Chatto and Windus, London, 1997).

Bowring, Finn. "A Lifeworld without a Subject: Habermas and the Pathologies of Modernity", in, Telos, No 106. (Winter, 1996, pp. 77-104).

Bunge, Mario. "Mechanism and Explanation", in, Philosophy of the Social Sciences, Volume 27, No.4. (December, 1997, pp. 410-465).

Burgess, Anthony. A Clockwork Orange, (W.W. Norton and Co., 1995).

Churchland, Paul. "Eliminative Materialism and Propositional Attitudes", in, Journal of Philosophy, Volume 78. (February, 1981).

Darwin, Charles. The Origin of Species, ed., Gillian Beer, (Oxford University Press, Oxford, 1996).

Dawkins, Richard. The Selfish Gene, (Oxford University Press, Oxford. 1976).

Descartes, Rene. Meditations on First Philosophy, trans., Lawrence J. Lafleur, (Liberal Arts Press, New York, 1961).

Dilthey, W. ed., H.P Rickman, Selected Writings, (Cambridge University Press, Cambridge, 1976).

Douglas, Mary. Implicit Meanings, (Routledge and Kegan Paul, London, 1975).

Eibl-Eibesfeldt, Irenaus. trans., Erich Mosbacher, The Biology of Peace and War, (Thames and Hudson, 1979).

Engels, Friedrich. Socialism: Utopian and Scientific, (Internet Archive, 2003).

Farganis, James. and Rousseas, Stephen W. "American Politics and the End of Ideology", in, British Journal of Sociology, Vol 14, No.4, (December, 1963, pp 347-362).

Fodor, Jerry. Psychosemantics, (The MIT Press, 1987).

Fukuyama, Francis. The End of History and the Last Man, (Hamish Hamilton, London, 1992).

Gadamer, Hans-Georg. Truth and Method, (Continuum, New York. 1975).

German, Tasmin C. and Wertz, Annie E. "Belief/Desire Reasoning in the Explanation of Behaviour: Do Actions Speak Louder than Words?", in, Cognition, Volume 105. (Issue 1, October, 2007).

Gooch, Stan. Total Man: Notes towards an Evolutionary Theory of Personality, (Sphere Books Ltd., London, 1975).

Gouldner, Alvin W. The Dialectic of Ideology and Technology, (Oxford University Press, Oxford. 1976).

Grice, Paul. "Logic and Conversation", in ed., P. Cole and J. Morgan. Syntax and Semantics, Volume 9, Pragmatics, (Academic Press, New York, 1975).

Habermas, Jurgen. Knowledge and Human Interests. trans., Jeremy J. Shapiro. (Beacon Press, Boston, 1971).

Legitimation Crisis. trans., Thomas McCarthy. (Beacon Press, Boston, 1975).

Communication and the Evolution of Society. trans., Thomas McCarthy. (Beacon Press, Boston, 1979).

The Theory of Communicative Action, Volume 1. trans., Thomas McCarthy. (Heinemann, London, 1984).

The Theory of Communicative Action, Volume 2. trans., Thomas McCarthy. (Polity Press, Cambridge, 1987).

On the Logic of the Social Sciences, trans., Shierry Nicholsen and Jerry A. Stark, (Polity Press, Cambridge, 1990).

Between Facts and Norms. trans., William Rehg. (Polity Press. 1998).

Harrison, Ross. On What there Must Be, (Oxford, 1973).

Harvey, David. The Limits to Capital, (University of Chicago Press, 1982).

Hegel G.W.F. The Phenomenology of Spirit, trans., A.V. Miller, (Oxford University Press, 1977).

Heidegger, Martin. Being and Time, trans., John MacQuarrie and Edward Robinson, (Harper and Row, New York, 1962)

   The Metaphysical Foundations of Logic, trans., Michael Heim, (Indiana University Press, Bloomington, 1984).

Hobbes, Thomas. Leviathan, (Penguin Books, London, 1968).

Hollingdale, R.J. ed., A Nietzsche Reader, (Penguin Books, London, 1977).

Horwich, Paul. Truth, (Blackwells, Oxford, 1990).

Hume, David. A Treatise of Human Nature, ed., L.A. Selby-Bigge, (Clarendon Press, Oxford, 1988).

Huxley, Aldous. Brave New World, (Marshall Cavendish, London, 1988).

Jaynes, Julian. The Origin of Consciousness in the Breakdown of the Bicameral Mind, (Mariner Books, 2000).

Kahneman, Daniel. Thinking, Fast and Slow, (Penguin Books, 2012).

Kant, Immanuel. Critique of Pure Reason, trans., Norman Kemp Smith, (St. Martins Press, New York, 1965).

   Critique of Practical Reason. trans., T. K. Abbott, (Prometheus Books, Amherst, 1996).

Kuhn, Thomas. The Structure of Scientific Revolutions, (University of Chicago Press, 1970).

Lawson, Hilary. The Post-modern Predicament, (Hutchison, London, 1985).

Lehrer, Keith. Theory of Knowledge, (Routledge, London, 1990).

Levinas, E. trans., Alphonso Lingus, Collected Philosophical Papers, (Martinus Nijhoff, Lancaster, 1987).

Locke, John. ed., John Yolton, An Essay Concerning Human Understanding, (Dent, London, 1977).

Lockwood, David. "Social Integration and System Integration", in ed., G.K. Zollschan and W. Hirsch, Explorations in Social Change, (Routledge, London, 1964).

Lyotard, Jean-Francois. "The Post-modern Condition", in ed., Kenneth Baynes, James Bohman, Thomas McCarthy, After Philosophy, (MIT Press, Boston, 1987. pp. 73-94).

Marshall, Michael. "Hothouse Earth is in the Horizon," in, New Scientist, No. 2839, Vol. 212. (19 November 2011. pp. 10-11).

Marx, Karl. Philosophic and Economic Manuscripts of 1844, (Internet Archive. 2003).

Mautner, Thomas. ed., Penguin Dictionary of Philosophy, (Penguin Books, London, 2000).

Mitchell, Melanie. and Newman, Mark. "Complex Systems Theory and Evolution", in, ed., M. Pagel Encyclopaedia of Evolution, (Oxford University Press, New York, 2002).

Mortimer, Chris. A Guide to Reflexive Therapy, (Lulu.com. 2013).

Mouzelis, Nicos. "Social and System Integration: Lockwood, Habermas, and Giddens", in, Sociology, Volume 31, No.1. (February, 1997. pp. 111-119).

Nietzsche, Friedrich. The Will to Power, trans., Walter Kauffman and R.J. Hollingdale, (Vintage Books, New York, 1967).

O'Neill, John. "Marxism and the Two Sciences", in, Philosophy of Social Sciences, Volume 11. (September, 1981).

Orwell, George. 1984, (Harper Collins, Canada. 2013).

Parsons, Talcott. ed., Leon Mayhew, On Institutions and Social Evolution, (University of Chicago Press, 1958).

Plange, Nii – K. Science and Social Theory, (South Pacific Review Press, Suva, Fiji. 1984).

Popper, Karl. Objective Knowledge, (Clarendon Press, Oxford, 1972).

"Evolutionary Epistemology," in, ed., J. W. Pollard, Evolutionary Theory: Paths into the Future, (John Wiley & Sons Ltd., London, 1984).

Quine, W.V.O. "On What There Is", in, From a Logical Point of View, (Harper and Row, New York. 1953).

Radnitzky, G. and Bartley, W. W. eds., Evolutionary Epistemology, Rationality, and the Sociology of Knowledge, (Open Court, 1993).

Ravetz, Jerome R. Scientific Knowledge and its Social Problems. Clarendon Press, Oxford. 1971.

Rees, Martin, J. Our Final Hour, (Basic Books, New York, 2003).

Rousseas, Stephen. Capitalism and Catastrophe, (Cambridge University Press, 1982).

Ryan, Alan. The Philosophy of the Social Sciences, (The Macmillan Press, London, 1970. pp 135-145).

Sartre, Jean-Paul. The Transcendence of the Ego: An Existentialist Theory of Consciousness, (Noonday Press, New York, 1957).

Schutz, Alfred. The Phenomenology of the Social World, trans., George Walsh and Frederick Lehnert, (Heinemann International, London, 1980).

Standing, Herbert F. Spirit in Evolution, (George Allen and Unwin, London, 1930).

Tilley, Christopher. Y. Miller, Daniel. Rowlands, M.T. Domination and Resistance, (Routledge, London, 1995).

Thompson. E.P. "Notes on Exterminism, the Last Stage of Civilization", in, Exterminism and the Cold War, (New Left Books, ed., Verso, London, 1982).

Wellman, Henry M. The Child's Theory of Mind, (The MIT Press, Boston, 1992).

Winch, Peter. The Idea of a Social Science, (Routledge and Kegan Paul, London, 1958).

Wolf, Susan. Freedom within Reason, (Oxford University Press. 1990).

Wood, Alex L, Merrett, Geoff V, Gunn, Steve R, Al-Hashimi, Bashir M, Shadbolt, Nigel R, Hall, Wendy. "Adaptive Sampling in Context-aware Systems: A Machine Learning Approach", in, IET Wireless Sensor Systems, (London, 2012).

GLOSSARY

Admittance realism: The concept that science creates paradigms, ways of looking at what-is, and has ways of experimentally testing and evaluating 'the truth' – the workability – of these views; and that not anything is workable because how-What-can-be-can-be-what-is, reality, does not admit of being known in any way. This is the concept of that, through creating and experimentally testing paradigms, science is succeeding or not with how it is possible to know how-What-can-be-can-be-what-is, that is, with how It admits (to us) of being known, something we do anyway routinely in finding and living how It can occur as us. We find (notice or not) that not anything can be made to work, the reason for this being reality; but this does not mean that aspects of how-What-can-be-can-be-what-is, of 'the Reason', cannot vary or alter with time or place. This may mean that, at times, unchanging general laws are being known, or that with time and evolving paradigms, the same thing is being known more accurately, or that different paradigms – in general ways of looking – may uncover different or seemingly contradictory properties, or that there is an also infinite truth.

Communicative sociation: Communicative sociation occurs in and amongst participants. It uses and creates ways what-is is experientioned. It is the practise, the being-process, in which social-integration occurs, and, in the conversation of self with itself, as self-communication, in which personal-integration occurs. To this degree, there is a need for communicative sociation in the conditions of participant-competence (see below). In communicative sociation, subjects participate in patterns, in a self-producing meaning (conversation, narrative-theme-involvement, satisficing), that can reach, and to a degree has to manage reaching, beyond immediate interpellation (see below) in the patterns of added system/B-engineered. In general, it involves the characteristics of the demands and opportunities in being society and person in systems that have created us, that operate in us, and that we add to our world and transform. In its reflexive relation with itself, communicative sociation monitors the effects on itself of interpellation in added systems, so evaluating, orienting to transforming, developing, avoiding, and the like. As expedient and preferred, involving or being about relations to, or use of, and reflexive coordination of feedback into, already added systems/B-engineered, it coordinates the clarification of issues such as of prudence or caution, along with evaluations. Ultimately, through the reach of communicative sociation, even the added language-system or legal-system or whatever it is can be reflexively examined and developed according to value. It can proceed as a consensus integration, or involve issues of influence or persuasion or domination.

Culture: A subsystem amongst others in the process of meta-system-(with-a-difference). Culture is the product of society-in-process, of narrative-theme-involved unitary mechanisms coordinating rationality-goals through concern, in combination with how the environment of added systems-B/engineered shapes activity. It consists in, and generates and uses, languages, tools, rules, added systems in general, conversations, and so on. It is the ways the being-processes (of subjects) are participants in the production-use of a world-of-understanding, involving creating and interpellation in and transformation of added systems, forms of self-regulation involved in being able to coordinate to fulfil roles, kinds of pursuits undertaken as ends-in-themselves, the differentiation of polity, and so on.

Drive-logic: Drive-logic is structured by the way the organizational structures of meta-systems, paradigms, interpellate patterns for the being-processes of subjects, and in processes generally. Patterning occurs through the demands of integration in added-system-B/engineered. Patterns reproduce and reinforce the interpellating-environment ('the box'), and so its limits on what a meta-system-process can be and become, on such as direction. So exterminist drive-logic is structured in that nation-states exist and are mobilized in domination/resistance-dynamics about economy 'achieving more to be better at war'. So the drive-logic mediated in capitalist organizational structure is a mobilization through economic activity to achieving profit. And so on. Issues in the recognition, development, and transformation of drive-logic feature in the management of survival-rationality.

Epistemologization-ontologization: The moment of 'knowing-of-being' that one becomes involved in in experiention-of-being (see experiention, experiention-of-being, below). One encounters, through reflexive meaning (in explicating 'the reflexive circle'), the issue of demonstrating the truth in the idea, in the experientioning of the meaning, that one exists. So, one finds oneself needing to develop a way, an 'epistemologized-rationality', for having and giving the reasons that demonstrate this truth, the truth in the assertion, namely, 'I know I am', involving an account of what it is to know (epistemology) and an account of 'that one is a being-process' (ontology, what there is).

Experiention (also see Experiention-of-being): Experiention is a new, compound word made up of experience and intention. Experiention is what 'experiences and intends', the existence of being that experiences and intends (consciousness). Experiention is being-process that cannot be seen externally but only lived and reflexively recognized during 'experiention-of-being'. In experiention-of-being experiention experientions itself occurring, we reflexively demonstrate to ourselves that and how we know that we are in the way we can. Experiention, we see, is there, us, what we are, before, in being a self-relation in it, we begin reflexively to develop knowledge of being it, and structure in it, involve in issues of there being

48

intending going on subconsciously and unconsciously, and neurophysiology and so forth. For us, instances of unitary mechanism, every experience involves intention, it in fact being insufficient to conceive experience without intention. That experiention 'relates' and self-relates, experientions itself 'thinking' and 'determining for itself ' and 'involved in contextual nuance', shapes a person's being-process, and involves issues in the foundations of human science.

Experiention-of-being: Experiention experientioning itself occurring, seeing that and what it is, eventually reconstructing 'unitary mechanism' (see Unitary Mechanism). In this, one becomes involved in 'the reflexive circle'. In experiention-of-being, one becomes involved, that is, with explicating the general issue of the being of what one knows of. In experiention-of-being, the issue is to demonstrate reproducibly that one is knowing of oneself being (see epistemologization-ontologization): one demonstrates the truth in the idea that one knows one is, by reflexively giving reasons to support it, by finding its truth for example in that it is necessarily predicated in the experiention of the issue of it.

Functional interpellation: The phenomenon of a role, of a performance of a patterned function, being demanded of (in) the subject's being-process by the functioning of the mechanisms and moments generally of which organizational structure, and the 'interpellating environment' (see below) of a situation of meta-system generally, is comprised. Interpellation (see below) is a term developed from Louis Althusser's use of the term to refer to states others interpellate one in. See, for example: [Althusser, Louis. 1972.]. Generally, functional interpellation is about the participant's experience in system-integrations (see below) involved in maintaining and reproducing systems-B/engineered: a player for example is functionally interpellated in the rules of a game, a machine-operator is functionally interpellated by the way the machine s/he is operating works, a shopper is functionally interpellated in the issue of purchase. A participant-competence is functionally interpellated in the demands of system-integration. Participating in a social-integration (see below), and in what realizing a social-integration is a condition in (for example, in aspects of being able to coordinate to manning a systems-B/engineered, or to creating a system, or to a transformation), are required moments in the conditions of participant competence, but there is no comparable interpellation-effect in the realization of social-integration, in the practise of communicative sociation, itself, it being a self-generating conversation.

Ground(s): The meaning a subject is in. The conditions of existence that make up the world of a subject. Different grounds, a chair or a path for example, involve one in different ways: in sitting in a chair and walking along a path, including such as the room the chair is in if it is in one, and what surrounds or is on the path. Sitting itself is a ground. Due to grounds person X's world is as it is relative to person Y's (not as well, better, different, whatever).

Holistification-adaptation: The idea of the 'ecological revolution', of that 'it can get/be it', of that human history in evolution, the dynamics of the reflexive loop, can succeed sufficiently holistically as far as the ecology is concerned, with the rationality of the ecological imperative, that economies are adapted within their ecological conditions and prevented from causing the collapse of the Earth's ecosystems. The idea that the reflexive loop can reach beyond the compulsions of drive-logics and the rest that drives economic activity outside of its ecological conditions and to ecocide and potentially extinction as a result of its not feeding-back sufficiently holistically as far as ecology is concerned, resulting in the collapse of the Earth's ecosystems. The idea that, through reflexive intervention in themselves via the reflexive loop, economies can become, can make themselves become, sufficiently holistic in orientation that things work – integrate – beyond the compulsions of interpellation in drive-logics and the rest that drives economic activity outside of its ecological conditions, that is, that drives it to ecocide and potentially extinction by not feeding back sufficiently holistically as far as ecology is concerned and as a result causing the collapse of the Earth's ecosystems. Knowledge useful in learning of ecological parameters includes such, for example, as: [Marshall, Michael. 2011]. Issues include, for example, biomass. Clearly, there is concomitantly a continuing issue of defending the conditions of human life from planetary and environmental conditions generally, including, in the long term, such as the sun exploding and galaxy-collisions.

Interpellation: Interpellation is a term developed from Louis Althusser's use of the term to refer to states others interpellate one in. See, for example: [Althusser, Louis. 1972.]. Here, it refers to the way properties of grounds subjects are in require patterns in the being-process of subjects participating in, who are in, them. So, functional interpellation (see above) is the phenomenon of a role (of a performance of function) being demanded of (in) the subject's being-process by the functioning of the mechanisms and moments of the interpellating-environment (see below). The interpellating-environment pre-structures possibilities-for-being, patterns in being-processes, including through the demands on the role of the subject in making system-integrations happen. So, a player, for example, is functionally interpellated in the rules of a game, a machine-operator functionally interpellated by the way the machine s/he is operating works, a shopper functionally interpellated in the issue of purchase, and so on. In general, a participant-competence is functionally interpellated in the demands of system-integration. Participating in a social-integration (see below), and in what realizing a social-integration is a condition in (for example, in aspects of being able to coordinate to manning a systems-B/engineered, or to creating a system, or to a transformation), are required moments in the conditions of participant competence, but there is no comparable interpellation-effect in the realization of social-integration, in the practise of communicative sociation, itself, it being a self-generating conversation.

Interpellating-environment: The environment which pre-structures possibilities-for- being, patterns in being-processes, including through the demands on the role of the subject in making system-integrations happen.

Logamic: Logic of drive-dynamic. Logically structured direction for or in (a) dynamic. So, for example, the drive-dynamics in an organizational structure drives exterminism, manifests a logic of direction for the whole, in the life of the mutually hostile meta-systems-with-a-difference involved. And so, without adapting the logamic in them, involving sufficient moments of consensus-integration (and including issues such as drawing resources from beyond the Earth), competing economies in an exterminist logamic are driven to expand beyond or otherwise threaten the ecological limits of their environment. And so, in its reflexive loop, reflexive-subject-in-grounds has a survival-issue of how to contain the exterminist logamic, or how to get it out of its grounds (for example by re-integrating in the logamic of paradigms/meta-paradigm). This is a dimension which, at a level, in at least the custom of ensuring defensive capabilities whilst concomitantly maintaining the principle of a non-hostile orientation to others, involves the continuing uncertainty of unknown potential threats, and so ultimately perhaps, at least in this way, is a dimension in the reflexive loop until the survival-system is perfected, if indeed, and particularly in the context of possible threat from the creative activities of the hostility of enemies, and infinity, it can ever ultimately be completely known to be so.

Meaning: In developing the definition of human existence as occurs here, a generalizable concept of meaning as 'what-is experientioned in a narrative-theme-mediated way', ultimately as 'what-is', is required: hence, understanding experientions meaning, is what experientions what-is in a narrative-theme-mediated way. What we, an understanding, experiention, is meaning, whether it is a feeling, or self-movement in being embodied, or a symbolically mediated abstract logical issue, or something incomprehensible. And for us, a meaning-relation is intrinsically a ground-relation and a value-relation. Grounds we are in are meaning in us and an issue of value, though not all of it may be content in our experientioning at any moment. What we are in (including what of us cannot be otherwise), is always already issue of what it is in world being preferred world. Much of meaning (reflexive meaning) is clear to us through the referent in the sign being in our experiention of ourselves occurring: meaningfulness is found in the internal self-relation of the signified, that is, in what experiention's relation to itself as experientioner. There is 'experientioned meaning', and there may be what of what-is noone will ever experiention.

Meanome/Meme: The meanome is the necessary experientioning, the combination of memes, in the conditions of there being a process of the production-use of a world-of-understanding at all. A meme is an organization of meaning that works in the process of the existence and reproduction of culture, of experientioned world, like a gene does in the process of the reproduction of the physiology involved. Meme is a concept introduced by Richard Dawkins in his The Selfish Gene: [Dawkins, Richard. 1976].

Meta-paradigm: The institution integrating paradigms (of meta-system), and consisting of a consensus amongst paradigms concerning parameters of mutual tolerance.

Meta-system: A meta-system is an instance of the whole of the process of life adapted in the production-use of the world-of-understanding (see below), of human life. Involving a difference, it is the whole that is made up of the functioning sub-systems (see below). It includes the difference, in unitary mechanism, in the participant competence subsystem-with-a-difference, of responsibility accompanied by the issue of freedom and contingency.

Meta-system integration: The coordination of all the moments required in a state of the meta-system of the production-use of a world-of-understanding effectively reproducing itself through time. The coordination of system- and social- integration is required. In the long term, as meta-system develops, ecological conditions are encountered, prompting moves beyond planets. The completion of meta-system integration is the perfection of the survival-system.

Narrative-theme: Element of transcendent structure intrinsically structuring our being-process. To be what it is, unitary mechanism (see below) must be participating in kinds of meaning-involvement. A narrative-theme is a kind of meaning-involvement, a meaning-aspect intrinsic in how what we are has to be concerned in and with the world if and as it is able to produce-use meaning – satisfice (see below) – and participate sufficiently competently (via coordination of rationality-goals) in production-use of world-of-understanding. Subjects, experientioning, we have to involve in narrative-themes if and as we understand and participate: if understanding, in experientioning what-is and being (also mutually) selected on as participant-competence (see below), what experientions participates in being comprehensible, truth-congruence, an issue of compatibility, an attitude and orientation, and an issue of emotional and spiritual stability generally and embodiment. So, the five narrative-themes participants must involve in, and co-coordinate, as a condition in sufficient competence, can be reflexively reconstructed and mutually recognized by us to be: 1) being comprehensible, 2) an issue of truth-congruence, 3) an issue of compatibility, 4) an attitude and orientation, and 5) an issue of emotional and spiritual stability generally and embodiment. What-is is what it is (is ground-relation, preferred world, etc.) for us, is value, in terms of these meaning-aspects, through a dynamic integration of these intrinsic, required, inter-linked being-process dimensions. These meaning-aspects, these ways the world-relation and process-story, what it can be to participate, is structured, shape discourses, and the differentiation of rules and institutions, and micro-dynamics and aspects of subsystems. So languages and issues of 'validity' and 'legitimacy', for example, can differentiate, and be explicitly constructed and maintained (in, for example, culture and polity), in the way the satisficing any participant practises includes comprehensibility- and compatibility-issues (concomitantly along with all the other aspects) for others, and so issues such as of the mutual recognition and practise of intersubjectively binding norms, ways, custom, and so on, and transformation.

Norm(s): For a subject, for unitary mechanism, every meaning (experientioning of what-is) is also encounter with value, that is, intrinsically involves its moment in the issue of world being preferred world. Experientioning's encounter with what-is is therefore always also normative, any meaning, supposed 'fact' or 'norm', also always being a moment in the field of relative value, a moment involving compatibility issues whose management intrinsically brings with it issues of 'meaning as norm'. Because both an external object, and a relation within self or with another person, involve issues of value, relations are all also normative relations, relations in which norms, differences regulating in encounter with value, enter in. Because content at all, meaning, is concomitantly organized, lived, 'also as value', 'also as issue of preferred ground', then, it intrinsically involves, is an issue of being, a norm; that is, it is something in an issue of preferred world, and so an evaluated way of being involving an issue of power in making it work, a satisficed moment that includes issues of realization, repression, and mutual recognition; and the subject, the evaluator, is 'transcendent normativity', i.e. is an ongoing carrier (involving being regulator) of issue of value in narrative-theme-mediated relation to what-is, involving management of social- and system-integration.

For subjects, 'facts' always already include normative issues, both through the regulation and organization of meanings involved in the conditions of their recognition as facts, and in the relation of regulating what, through being recognized as such, as facts, they otherwise mean in world being preferred world. For a subject meaning is concomitantly a personally and mutually practiced – satisficed – norm, and this is because relations of encounter with value are intrinsic in any meaning and require that, in being a being-process, through concern, the subject regulate itself in, as regards, them. For subjects 'values' express attitudes in the way meanings are value, shaping motivations in making world preferred world (for example a world in which facts are sought and known), including in institutionalizing mutually accepted rules and conditions – 'norms' – of conduct. (See Satisfice.)

Paradigm: An actually lived model for organizational structure (or ultimately scheme, political, epistemological, whatever, at all) in the production-use of world-understanding. In participating in a meta-system, one is in the paradigm it is. In the meta-paradigm/paradigm's model, a paradigm consists of, is created by and held together through, a consensus amongst participants in it. (See, for example: [Kuhn, Thomas. 1970].)

Participant-competence: One of the 'subsystems' of meta-system. The subsystem(-with-a-difference) in which the process of the subject – unitary mechanism – achieves and supplies its role in the process it is what it is in, that is, in the conditions of there being a process of the production-use of a world-of-understanding (human life). Its doing this includes, for example, 1) its fulfilling the demands on it of system-integration, including innovating and transformation of systems-B/engineered, and maintenance of the reflexive loop (see below) generally, and 2) finding sufficient personal- and social-integration through communicative sociation that it can be sufficiently. The difference in unitary mechanism, in the participant competence subsystem-with-a-difference in its value/power-dynamic, is in that the experience of responsibility includes an experience of being self-causing, bringing with it issues of being caused in, and of freedom, and bringing contingency to the concept of 'laws of system-function'. Nonetheless, through working with knowledge of significant and relevant relations of environment, narrative-themes and neural pathways, and genes and memes, and so on, self-systematization in the conditions of development and competence, systematically criterionizing what counts, is in aspects possible for culture, and is amongst issues of uses of reflexive human science.

Rationality: The process of having and giving reasons in how meaning is being produced-used, and so in how we are a being-process. There are different ways of having and giving reasons: epistemologized ways as opposed to ways based on superstitions for example, ethically principled ways, purely passion-driven ways, ways that generalize themselves or not, scientific ways involving experimental evidence. There are different levels and aspects of it, some to do for example with beliefs we have about how things work that we have acquired through experience. Some ways for example follow how causes in ways objects are used work: we hit a ball with a bat because we have learned this is a way to move the ball; this is the reason we hit the ball with bat, the rationality we are involved in when we hit the ball, rationality that is refined in terms of issues of how we hit the ball, etc.

Reason: The way, through narrative-theme-involvement, evaluation is lived – satisficed and triadized – and world is made to be, norms and values occur, and so forth. In making satisficing and triadizing, a world, work, the subject is oriented in explaining to him/her self why s/he is doing something, or what is going on generally. In seeing, for example, that the reason (i.e. why, involving cause) a light-bulb illuminates, includes what happens when a switch is turned, a subject is seeing what s/he knows about why something happens, and encounters ways issues of relations to value are regulated, in turn therefore also encountering sources of – aspects of reason in – his or her activity: in normal circumstances the reason at work in turning a light-bulb switch involves preference about whether or not its light is wanted because it is recognized that turning a switch manifests or removes its light, which light (produced-used meaning) is wanted or not for reading or sleeping etc., that is, in making preferred world work in this way. We hit a ball with a bat because we learn and know this is how we can move the ball and because we want to move the ball; we hit the ball for these reasons, the preference to so hit the ball is a reason in why we so do. As we produce-use and form understanding, which can include epistemologized and non-epsitemologized aspects, we deduce, infer, get involved in syllogisms, value being able to switch on a light, or hit a ball, or find an ethics, and so forth. The subject explains to him/her self why s/he is doing something, or what is going on generally, by recognizing the reasons in and for – to – it. In seeing that the reason (why, involving cause) a light-bulb illuminates, includes the truth of what happens when a switch is turned, including the preference to switch it on, or accident if it happens accidentally, and the physics in the motion of electrons that occurs when this is done or occurs, a subject is seeing what s/he knows about why, the reasons in why, something happens, so encountering ways relations to value are structured and regulated, so encountering sources of – aspects of reason in – his/ her activity. So: the reason for turning a light-bulb switch involves preference about whether or not its light is wanted because turning a switch manifests or removes its light, which light (produced-used meaning) is wanted or not for reading or sleeping etc., that is, in making preferred world work in this way. So, in being what we are, in the conditions of making a world work, in being as we must if and as what we are in the life we are what we are in works, we become involved in epistemologized and non-epistemologized rationalities, ways of seeing and seeking and using reasons, in deploying mathematization, and so forth.

Reflexive/reflexivity/reflexivization: What we are experientioning itself occurring and becoming self-related through self-awareness, involving experientionof-being, self-reconstruction, self-application, reflexive human science, reflexive therapy, etc. Become reflexive, in reflexivity, our being-process is aware that it is, and is what it is related to itself, experientioning itself occurring and involved in self-reconstruction and self-application and producing its world as it prefers and can make work generally. Much of meaning (reflexive meaning) is clear to us through the referent in the sign being in our experiention of ourselves occurring: we evaluate, interpret, in general participate in production-use etc. Reflexivization proceeds upon the basis of knowledge of unitary mechanism, that is, knowledge of what is generalizable in 'what I am and how I work in the process of being what I am'; and reflexivity is one's knowing oneself as the functioning – with a difference – of this mechanism, and in this is knowing oneself to be this generalizable mechanism occurring in specifically personally relevant ways, existing in its spiritual environment, involved in its constructions, guided and influenced by its values and objectives, seeking objectivity, and so on. Reflexive, having reflexivized, our being-process exists in awareness of that and what it is, with a reflexive perspective on what it does; and, developing itself in these terms, it becomes able to be scientifically.

Reflexive human science: Science of us that we do via the evidence, through reflexive awareness, of that and what our being-process is. This is scientific awareness that includes our being us amongst its object domain. This science has reflexive-subject-in-grounds as its explanandum (grounds in which include physiology etc.). The truth of the assertions of the science are satisficed reflexively: what counts as evidence that the assertions are accurate is found in the mutual testing of the reproducible in what we experiention to be occurring when we experiention ourselves occurring.

Reflexive loop: The space in human life manifest through the becoming reflexive (reflexivization) of what we are (see Reflexive). Reflexively mediated feedback in the process of human life in which – in such as reflexive therapy (see below) as here explicated – we have reconstructed and recognized, and systematically-with-a-difference can consider, us and our activity in how the process of our life occurs.

Reflexive need: Reflexive need includes our need to know about our existence and self-relation, about that and what we are, and about our need to be and that we use ideas of that and what we are in being. It includes need we have to clarify ideas about that our being-process is what it is, and that being what we are we need to cause ourselves to be, and that as we do we become involved in ideas about the world and how we should be. These are ideas that we self-apply with, and about why self-application is as it is: for example, that we should manage truth and knowledge because not anything can be made to work and we need an idea of truth and knowledge to make things work, that we see that not any rules are wanted by those who mutually recognize one another as collectively and voluntarily subject to them, etc.

Reflexive-subject-in-grounds: Phrase denoting the subject's situation in the world as it can reconstruct it for itself, that is, it's existence as it is, as unitary mechanism in what it is in, which for it are grounds, including conditions of existence, and its own existence and potentials.

Reflexive therapy: Reflexive therapy addresses reflexive need. Reflexive need includes our need to know about our existence and self-relation, about that and what we are and cause ourselves to be and become involved in ideas about the world and how we should be, ideas about that we self-apply and why self-application is as it is (see Reflexive need). Reflexive therapy addresses reflexive need scientifically, through reflexive human science, developing ideas about what must be in that we are and can be as we are being, and the like: it achieves ideas of that and what we are scientifically, and uses them in self-application and being together, in general, in addressing and managing reflexive need.

Satisfice: To satisfice, which is concomitant with an issue of triadizing in the process of unitary mechanism, is to be a subject-involved-in-understanding working out and through the meaning (interpreting, believing, mutually establishing, desiring, self-regulating in its content-states, sublimating-utilizing, 'getting it', developing rationalities, and whatever else) it is involved in and occurs − is itself − as, including in regulating in and institutionalizing the norms of its being-process. To satisfice is to be, to exist producing-using meaning in an issue of the world being as preferred as can be made to work. Satisficing is a process of self-involvement and intersubjectivity, and happens in every moment − in the continuum − in which what-is becomes and is experientioned meaning for subjects. Satisficing happens if and as an experientioning, a being-process, is, and involves the five narrative-themes concomitantly. One could say: 'it' always already is satisficing if and as a being-process is occurring. Culture is the outcome of − and is reproduced through, exists as − 'a satisficing-process'. The term satisfice is also used in psychology: according to the Penguin dictionary of psychology the term is from Herb Simon, being used to signify acceptance of a judgement or choice as one that is good enough when optimal solutions are not to be had. In the usage here, a satisfice could be, but need not always be and often isn't, an 'optimal solution'. In general, there is when one has to make do with other than one's preferred ground as such, the value triadized being to do the best that one can make work in the circumstances.

Science: Activity of achieving truth-congruence (knowledge) in a way that involves recognition of such as reproducibility, mechanisms, structures and laws in how What-can-be-is-what-is (in Truth). It involves issues of ways of establishing truth-congruence (method, ontology), discourse and consensus concerning evidence of truth-congruence, admittance realism, and so on. It is not exclusively an issue of relating systems of abstract concepts to empirical (that is, external) data, but is also reflexive. (See Reflexive human science.)

Social-integration: The structural demand, in the process of a meta-system, that individuals (participants, subjects) be sufficiently mutually coordinated in communicative sociation (see Communicative sociation), that a process of culture, that the process of the unchanging structure of self-coordinating, narrative-theme- and systems-B/engineered-involved, interacting intersubjectivity and participant competence, be able to exist (see Society). Communicative sociation is significant in the conditions of participant-competence, society, and culture; and, in social-integration, it is a necessary condition in meta-system-integration along with sufficient system-integration. Sufficient social-integration is linked with and enables aspects of personal-integration, development, polity, and self- and mutual-regulation of capacity to man added systems, to reach and reproduce, and feedback into and transform, the interpellations of system-integration (see System-integration.) In together with others, each needs ways, for example, of sufficiently managing 'what another means in, the difference another makes to, his or her state' and so on, so ensuring the stability of interaction and society. Mechanisms can be and are mutually deployed that have significance in managing social-integration, for example in constructing and enforcing a notion of legitimacy in satisficing, but the moments of social-integration itself, including, for example, such as issues of repression within participants, is apart from and alongside moments of functional interpellation; if it were not then it would be nothing more than a mode of system-integration. Trends in culture, in how society is a process, in for example how involvement in added systems-B/engineered is occurring, can threaten the stability of the social structure, that is, ultimately, the conditions of there being the communicative sociation of narrative-theme-involved intersubjectivity and participant competence at all, this dimension involving participants in ongoing issues of regulating culture so as to ensure society, and so too that it, a working culture itself, consisting of both subject and system, is possible.

Society: A subsystem-with-a-difference interlinked with the other subsystems of meta-system. Society is the unchanging structure of narrative-theme-mediated interaction proceeding through the communicative sociation of subjects. Society is an unchanging structure whose process, consisting of subjects' satisficing, is culture.

It is culture which can involve change and that admits of different versions. The integration of society, the stability of the structure through time, occurs through the practise of participants, in the mutual coordination of communicative sociation (see Social integration), in the production-use of culture generally. The production-use of culture generally results in the differentiation of polity and economy, and issues of how added systems-B/engineered (see below) affect the conditions of society, participant competence, and culture generally emerge in history and evolution.

Subject/subject-form: What participates, the individuated and intersubjectively linked experientioning-of-meaning-process that, for itself, produces-uses meaning to make its world as it prefers as it can make work, and that has an issue of participant competence, that is, that is structured in unitary mechanism (see Unitary mechanism) and self-coordinates. This includes its relations and obligations in intersubjectivity, its issues of coordinating value, of mutual recognition and nurture and so on, with others, including, for example, its mutual constructions, in satisficing (see Satisficing), of such as notions of legitimacy, regime, and so on. Unitary mechanism operates, and its subjectivity is the emergent presence, ego, one or another degree of coordinated understanding and self-relation, of conditioned and interpellated experientioning, identity, etc., in general subject-form, in being-process. The subject is in a world of relations and objects, of subjectivity and objectivity, of relations to itself, to others, and to the rest of It (What-can-be-being-what-is) generally, including, reflexively, in experientioning itself occurring and being itself, in making itself 'object' for itself in reflexive human science (see Reflexive human science). Reflexivized (see Reflexive), the subject has clarified its 'reflexive need' (see Reflexive need) and 'reflexive self-presence whatever the content', has reconstructed that it is a being-process structured (definitively) by unitary mechanism (see Unitary mechanism), and that it exists as 'reflexive-subject-in-grounds' (see Reflexive-subject-in-grounds) and in subject-forms. At a level, the subject encounters itself as an adaptive strategy in a life, a strategy that exists, that experientions being and that it operates as unitary mechanism and as shaped into subject-forms, that it self-causes and can reflexively self-apply and, at a level, be scientifically and manage self-evolution.

Sublimation-utilization: 'What we do with instinct'. The subject's finding and regulating, creating and converting, of energy through participating (satisficing) in aspects of how it's being-process works and can be made to be. In a being-process energy happens and can and needs to be transformed from one form into another: for example, digestion, or a desire for food or sex, into an application of concentration, a light, a 'the that'. It happens, and finding and regulating, creating and converting, energy occurs in relations with self and others, and in relations to objects and environments (cathexis, libido); that is, it occurs in and in relation to grounds generally, including embodiment, narratives and fantasy, quantity and quality, indulgence and abstinence. It can involve issues of repression and morals, includes connecting and sharing, creating and happening together, and synergies. Transforming can be about subjects inhibiting and transforming a form of energy, or utilizing it – what happens in interaction for example – more directly, without mediating inhibits; and subjects can work the two together. Inhibiting or conversion or utilization, what happens intersubjectively, turns, are grounds with consequences in how being – 'times', innovation, individual/social coordination, 'what-about-it-happens-and-does', thought, state, health, direction – proceeds and happens. Aesthetic and discipline make a difference in it.

Subsystem: The sub-systems, with-a-difference, of meta-system, include: society, culture, participant-competence, economy/ecology, and polity. Together these make up the meta-system of the process of life producing-using a world-of-understanding. There is tension in the process of human existence being analysed exclusively in the terms of a systems-theory which sees that the moves of a system can be predicted via knowledge of its situation and laws of functioning, however. It is a tension rooted in the property: 'responsibility with an issue of freedom' in the subject (in unitary mechanism). This property includes the issue of what, if anything else, is causing in how what experientions itself, also, to be a self-causative self-activity, operates. Because there is an issue of system-with-a-difference in participant competence (see above), and because subsystems are interconnected, this tension is manifest in issues of contingency in transforming and adding systems. And this tension is apparent in such as the dialectic of communicative sociation and functional integration. In reflexive human science generally, the 'with-a-difference' issue means that human life, seen as a process whose laws include the law of unitary mechanism, and laws surrounding the demands of narrative-theme-involvement and the properties of the field of relative value, involves variables, issues of contingency and flexibility, through subjectivity (value/power, communicative sociation and through it conceptual development, and so on) and subject-led feedback, innovation and transformation generally. So some, such as Jurgen Habermas, for example, speak of 'system' and 'lifeworld': [Habermas, Jurgen. 1984, 1987].

System: The interrelated elements that cohere into a self-maintaining whole, and that draw resources from an environment. Ideally, the moves of a system can be predicted via knowledge of its situation and laws of functioning. 'The unexpected' and learning aside, for this reason it is not possible to completely label human life a system without (via reflexivization) acknowledging the issue of the difference in the subject: though much is structured and approachable as a functioning system, there is an issue of freedom and contingency in a mechanism that experientions responsibility and is involved in value/power, that is, in one of the elements, in 'mechanism as freedom'. So some, such as Jurgen Habermas, for example, speak of 'system' and 'lifeworld': [Habermas, Jurgen. 1984, 1987].

Systems-A/lifeworld: The systems within existence, evolution and whatever else is involved, along with the subject-practices and its spirit-environment, which are the primordial conditions in there being a process of the production-use of a world-of-understanding, of a meta-system-with-a-difference of human life, at all. These systems include, for example, the vision and digestive systems, and the planetary system. And this includes the lifeworld of practicing subjectivity that emerges in the process of these systems. So, with the emergent phenomena of social-integration, cultural reproduction, and participant-competence, phenomena emergent concomitantly with unitary mechanism, with the practises of the subjectivity, the experientioning entity, in terms of which a process is possible at all, a lifeworld manifests: systems-A/lifeworld begins, and as it operates systems-B/engineered are created and transformed. In such as Jurgen Habermas' contrasting of 'system and lifeworld' ([Habermas, Jurgen. 1984, 1987], following such as [Schutz, Alfred. 1980]), the dimension of system (eyes, language, voice-box) in the conditions of lifeworld, in contrast to added-systems-B/engineered, is unclear. Because it is here seen that a subject and culture are tied systematically into conditions of existence, conditions in it being able to exist personally, and socially integrate and create and reproduce culture, here issues of the 'colonization of the lifeworld' (Habermas' phrase) are seen to be about how 'added systems' – added functional integrations, added systems-B/engineered – interfere in 'lifeworld mechanisms, dynamics and process' and 'conditions of its happenings generally', including systems-A/lifeworld generally. This is about interference in the conditions of there being subjects doing what subjects have to be in and do, in the way of communicative sociation and whatever else, to be able to become and be competent participants and thereby fulfil the demands of the role of subjects in the conditions of there being person, society and cultural reproduction, in general, in there being satisficing-practise in the process of a life producing-using a world-of-understanding, a meta-system, at all. Because there are conditions that can be interfered in through 'colonization-moments' at all, there must be an issue at this level of systems-A/lifeworld, i.e. an issue of the mechanisms and whatever else that are conditions that can be interfered in in this way. One of these conditions is 'mechanism as freedom', that is, unitary mechanism, is our, the human, being-process at the level of the subject. This is a level in which issues are lived – experientioned, and experientioned reflexively – as issues, issues involving reflexive need, issues of greed and compulsion and responsibility and whatever else.

Systems-B/engineered: Systems the life involved in the production-use of the world-of-understanding, human life (us), adds to itself, in its process, as it proceeds with making the world it is in the world it prefers as it can make it work. Examples of added systems include spears, houses, cars, and nuclear power plants. (See Systems-A/ lifeworld.)

System-integration: System-integrations relevant to subjects are patterns that 'directly interpellate the being-process of subjects participating in them' (see Interpellation). These patterns, which are conditions in the functioning of the systems, shape the role subject-activity has in the reproduction of the systems, systems which, in turn, are grounds human life is in. For subjectivity, participating in the process of a species of life involved in producing-using a world-of-understanding includes meeting the demands of its role in reproducing system-integrations. The subject reproduces directly interpellating patterns involved in the reproduction of systems, and in general the subject ensures the subject-activity/role (competence) relevant in the coordination of these patterns, and this can involve social integrations as well. The personal- and social-integration of participants relevant in the maintenance of competence are a condition in system-integrations involving subject-activity being possible, but are not system-integrations as such. System-integrations occur along with social-integrations in the dialectic of communicative sociation (see Communicative sociation) and functional interpellation (see Functional interpellation) an operating unitary mechanism is in. Together, system- and social-integrations reproduce meta-system-integration, that is, the integration of meta-system-with-a-difference. For subjects, system-integration is a concept primarily about being in added systems-B/engineered, but it is about being in systems-A as well (see Systems-A/lifeworld): so, we know healthy living and medicine play a part in maintaining health. So, the maintenance, and correct use of and participation in, traffic lights are required in the traffic system working. Establishing system-integrated ways of doing things constructs and can stabilize aspects of the process, ordering and dampening the vagaries of 'the human difference': for example, creating a legal-system, and helping to ensure routines of maintenance and standards of safety. (Issues in both systems-A and systems-B, and of both a legal-system and maintenance and safety, feature, for example, in a smoking ban.) The risk in totalizing levels of system-integration, a situation in which all moments of subjectivity have become completely interpellated in the patterned functions of added mechanisms, is that other moments of subjectivity and culture become stifled by the interpellating patterned functions of engineered moments and routines, threatening, through such as 'colonization of the lifeworld' [Habermas, Jurgen. 1984, 1987] and drive-logic, unmanageable repression, the possibility of personal development and personal- and social-integration, and properties of polity and that are in people, including freedom, threatening ultimately aspects of identity and flexibility, feedback and adaptability, of integration and conditions of human being and life generally. Similarly, too much emphasis on social-integration (see above) at the exclusion of system-integration, can limit differentiation and aspects of the process of adding systems-B/engineered, and be impractical and over-demanding. Tendency to totalization of system-integrations is conceivable, for example, in every moment becoming in some way commoditized (involving removal of things that cannot be commoditized), or in everything being assumed 'already known' or 'a named right': it could involve tendencies such as 'applied systems for bringing up a child as competent participant', or 'applied systems for ensuring participant P does D', or 'applied systems for noticing, identifying, and correcting deviations from an established institutionalized being-process ('it')', or 'a system for doing it', or 'needing it totalitarian, or completely technocratic'. In general, as system-integrations, drive-logics, and so on, manifest in the historical and evolutionary process, 'reflexive monitoring' of such as motive, method, and product, including in some degree 'science of science', in a 'reflexive loop', is emergent potential, through reflexive human science, for 'reflexive-subject-in-grounds' (see Reflexive-subject-in-grounds), indeed, is intrinsic practical intent for as long as a critical subjectivity, a free, reflexivized, spirit that has not succumbed to 'total interpellation', exists reflexively in its relation to grounds, involving issues of reflexive subjectivity freely selecting its grounds (choosing which applied systems it wants to be in), and issue of how paradigms are differentiated and select on one another: for example, whether through domination and so on. In a perfected survival-system, involving overcoming such as the sun exploding, or galaxies colliding, or black holes being threats to survival, it seems logical that the activities of society are ideally ultimately freed from any need for a role in coordinating the systems survival depends on.

'The that': The inner energy and light, life-energies and qualities, that are there for the subject and in intersubjectivity from within. These occur in the happening of the process, depending also on grounds, though that are not necessarily noticed or oriented to, lived, by the subject. These moments, involving telergies, pressures, tendencies, connectivity, exchanges, timings, and so on, − it -, can be manipulated in. How 'the that' happens is sensitive to conjunctions of meaning in grounds and natures, congruence and incongruence, style, surfaces, space, manners, custom, architecture, aesthetic generally, and so on.

Triadizing: A subject triadizes value when it turns possibility into achieved difference via actualization of the possibility. Triadizing is intrinsic in value-realization in the process of unitary mechanism. Triadizing is concomitant with satisficing as meaning is produced-used, as understanding occurs and world is made as preferred as can be made to work. To make a cup of tea, for example, is to triadize a value, as is to think thought; and actualizing these possibilities concomitantly involves satisficing relevant meanings (such as sourcing the tea, managing compatibility issues in that tea is being made, finding words if any are relevant, and so forth). As a being-process moves through its triadizations of value (getting up, making tea, thinking through what is going on, what to do and how to do it, and so on), being through the day, proceeding from the first waking moment to the moment of return to unconscious sleep, the boundaries between triadizations merge, and some contain others: the day consists in triadizations of value, some triads containing sub-triads and so on. In a sense, a human lifetime is the meta-triad, the actualization of death through life, which contains the triads-within-triads that a being-process is made up of. An issue of triadization is in satisficing, and vice versa: for example, amongst 'thought that just happens', and the rest in the functioning of what we are, we triadize in an issue of changing a situation when we arrive, through satisficing data, at the meaning that remaining in it is not the preferred option (involving interpretation and whatever else, and itself also a triadizing): we comprehend it, take an attitude towards it, become unstable in relation to its truth, and, finding it incompatible, in the value of wanting to be free from it, we do what can be made to work to change it. Triadizing this or any value includes satisficing in doing so. Repression occurs if we begin satisficing a value but cannot complete satisficing with the power to make the value actual, by actually triadizing it: then the dynamic 'satisficing value/triadizing' triadizes another, generally next best, thing that it can make work. The elements and levels in 'there being thought' and 'realizing and thinking thought', begin in such as 'thought that just occurs', and become more complex as we self-cause and coordinate it in our world generally. In a dynamic that is completing actualizations of value or not, our being-process, experientioning of what-is, is involved in the difference of the satisficing involved in an actualizing; and, in achieving satisficing at all, a level of 'the triad' intrinsic in how our being-process works is concomitantly occurring. Thinking thought T, in general being the experientioning, involved in doing a satisficing, we coordinate T's relevance in the triadizing it has a role in. And perhaps, in general, 'in the process of will-to-power-of- freedom', triadizing occurs in the activity of 'the Subject', 'Spirit- as-such'; and, as 'thoughts just occur', it includes other unconscious mechanisms as well.

Understanding: To understand is to experition what-is in a narrative-theme-mediated way. Understanding is what experitions meaning and, here, that reflexively reconstructs that, for it, this occurs dynamically, in practise, through the narrative-theme-involvement of subjectivity, that is, the narrative-theme-involvement of an experientioning entity producing-using meaning – experientioning what-is – to make the world as it prefers as it can make it work, a satisficing/triadizing entity.

Here, the concept of understanding is a concept of what experientions what-is in a narrative-theme-mediated way in its role satisficing and triadizing in a process that coordinates the production-use of meaning (of experiention of what-is) to make its world as it prefers as it can make it work, involving spirit, mind-brain, will, rationalities, a field of relative value, and whatever else, whatever other grounds. Here, then, understanding, experientioning what-is in a narrative-theme-mediated way, is seen to be as it is in being what it is in the conditions of there being the satisficing/triadizing of unitary mechanism. (See Satisficing, Triadizing, Unitary mechanism.)

Unitary mechanism: The mechanism structuring the subject's being-process according to which each human being produces-uses meaning to make the world as s/he prefers as s/he can make work. The concept of unitary mechanism is the concept of what structures and is definitive of the subject's being-process (see Subject), of what occurs if and as participation is. It is the concept: each human being-process exists producing-using meaning to make the world as s/he prefers as s/he can make work. In reflexive human science, this mechanism is reconstructed as what defines human life at the minimal level of the structure of a subject's being-process, the level of existence. Seeing and reconstructing unitary mechanism in reflexive human science (spirit science), reconstructs the fact of our existence, our being-process (experientioning), as the existence of a spirit-mechanism, a definitive level in a system-with-a-difference. This reconstruction, enabling the minimal definition of human life, indicates that in human science it is a reductionist error simply to 'cut to the physiology' because we, – reflexive-subject-in-grounds –, (also) exist, understanding, in and in relation to ourselves and meaning we experiention generally, including information about physiology and our relation to it. It seems likely, and is consistent with our experience, that our existing, a being-process experientioning itself occurring and applied to itself, is something in how our life is adapted, in its survival-strategy. Meaning is satisficed and triadized in what we are. For example, we experience self-causing to use knowledge of physiology to act on it according to value, to intervene in its ground as it makes a difference to us to do, to take a pain-killer to get relief from pain, for example. And, as such as out of body experientioning, co-perceiving, and so on, actually occurs, this existence seemingly reaches 'apart from the physiological ground' as well. This reconstruction indicates that we exist in operations of unitary mechanism, and exist there with causal relevance for the explanation of the activity of human life. It indicates this, as well, because manipulative intervention in this mechanism can work, because this mechanism can be manipulated in in persuasion and rhetoric, in advertising, propaganda, conditioning, and the like, in the formation of its meanome (see Meanome), in how it does what a participant, at least, has to be doing if and as a process of production-use is. It indicates this because manipulative planned patterning of functional integrations is a potential, along with uncoupling satisficing and triadizing from communicative sociation and chaining identity, and managing repression in the subject's response, if there is any, and the like. This is a mechanism in the operations of which there is an issue of what must be occurring if and as its role in the life it is what it is in is being fulfilled (narrative-theme-involvement, for example), a mechanism the examination of which is a beginning for a science of human life. And because, at a level, this mechanism defines human being, issues of de-humanization are apparent in terms of it.

Value: For subjects in the process of life producing-using a world-of-understanding, subjects that are structured in unitary mechanism, value is the driving issue of world being as preferred: a subject is always already confronted with the issue of its world being as it prefers as it can make it work. A subject's value is the subject's preference in relation to grounds, potential and actual. Preference is evaluated through reasons, in rationalities, evaluations that integrate issues such as function and aesthetic, and aesthetic function (for example as ground in 'the that', telergy, and so on). So, if both lamps work, subject S prefers lamp A to lamp B for aesthetic reasons, so, if it can be made to work, preferring to make lamp A a part of his/her world rather than lamp B. Subjects, structured in unitary mechanism, intrinsically see meaning as something in preferred world because experiention-of-what-is, what human understanding is, is what it is in a being-process for subjects in how life producing-using-a-world-of-understanding works: unitary mechanism operates, and its subjectivity, the emergent presence, ego, biography, one or another degree of coordinated understanding and self-relation, etc. in being-process (see Subject), coordinates its relation in value. Value is an intrinsic issue for subjects through how unitary mechanism works in conjunction with the issue of power, the issue of the capacity to realize value, meaning that in culture there is also intrinsic issue of relative value. And value is selected on by evolution. Value means that issues of relative value are intrinsic in and between individuals and cultures, and are so in how evolution works, including issues of domination/resistance, democracy, consensus, and so on, in how compatibility (relative value) issues are managed.

Value is present in experiention through concern, inspiring creativity, innovation and transformation, as well as preservation and conservation. Value inspires thought, is thought and emotion. Value manifests in the achieved differences a subject actualizes as it makes its world as it prefers, and is there anyway in the situation of relation to grounds whether or not preferred ground can actually be made to work, a situation involving issues of levels of repression, resistance to an order of things, and so on. Achieving a goal, such as satisfying a thirst, or overcoming an addiction, or organizing a way to know, or such as reaching a shared understanding with another, is value, just as grounds one is in anyway involve an issue of being value. An actual preferred path may not be the originally wanted one, or the most desired one, but the one selected because it can sensibly be made to work; the path selected may not in the end be the one that can be made to work, meaning that something else then becomes value; one may not want to be in a ground one is in, involving issues of repression-management. Forms of measurement of value emerge and manifest through externalizations, but often it exists only reflexively, intrinsic in the dynamic that is human existence.

Will-to-power-of-freedom: The process manifest (in evolution) through subjectivity, structured in unitary mechanism. Possibly a property, dynamic, in how What-can-be-is-What-is of which the subject is a level of expression, of manifestation, a property perhaps of 'Spirit', or the 'Subject', or 'World-spirit'.

World-of-understanding: The world, meaning, there, in history and evolution, through the operations of understanding, that is, through the operations of unitary mechanism coordinated through the subject's narrative-theme-mediated experientioning of what-is. Such meaning includes memory, buildings, languages, relics, and so on: it includes the spiritual and material environment, involving 'current ways being-processes occur (including in future-orientations)', and the way that, through understanding-oriented activity, the past has shaped and put – added – systems, many of which are external-material, in architecture, scientific method, and whatever else. The process of the world-of-understanding is the process of the world an understanding, the producing-using subject structured in unitary mechanism and manifest in varieties of subject-form, produces-uses as it experientions meaning and, through narrative-theme-involvement, develops rationalities, a relation to truth and ways of making things work, that is, fulfils its role in the life it is what it is in. A world-of-understanding intrinsically includes the issue and dynamic of value/power. Against the background of the minimal definition through unitary mechanism, the definition of human life can be developed in terms of the idea that it is 'life that is involved in producing-using a world-of-understanding', involving world-aspects such as the external, the internal, and the intersubjective, involving a meta-system, and so on. In human life, nature and spirit are involved in 'there being the process of a species of life involved in producing-using a world-of-understanding'. Human life is a meta-system-with-a-difference of the process of the production-use of a world-of-understanding. In the external world, for example, there is what is in the environment as a result of the activities of action oriented through understanding and that can be recognized to be 'a part of the world-of-understanding', or 'a part of the world, the external environment, understanding has created, designed and built etc., and is continuously involved in'. In the internal world there are the thought-processes and the rest in terms of which action-oriented understanding coordinates actions. Intersubjectively there is a field of relative value, narrative-theme-involved interaction. There are languages, organizational structures such as institutions that differentiate in the process of a life producing-using a world-of-understanding. And so on. (See Meta-system.)

INDEX

www.ingramcontent.com/pod-product-compliance
Lightning Source LLC
Chambersburg PA
CBHW071253170526
45165CB00003B/1322